Matheus V. D. Gueri

Anaerobic Digestion of Food Waste

Matheus V. D. Gueri

Anaerobic Digestion of Food Waste

Study with batch and semi-continuous reactors

ScienciaScripts

Imprint

Cover image: www.ingimage.com

This book is a translation from the original published under ISBN 978-613-9-61809-5.

Publisher:
Sciencia Scripts
is a trademark of
Dodo Books Indian Ocean Ltd. and OmniScriptum S.R.L publishing group

120 High Road, East Finchley, London, N2 9ED, United Kingdom
Str. Armeneasca 28/1, office 1, Chisinau MD-2012, Republic of Moldova, Europe
Printed at: see last page
ISBN: 978-620-7-70549-8

SUMMARY

SUMMARY 2
1. INTRODUCTION 3
2. Objectives 8
3. literature REVIEW 9
4. Materials and methods 35
5. Results and discussion 47
6. Conclusions 74
7. REFERENCES 76

evaluation of the anaerobic digestion process of food waste in reactors

BATCH AND SEMI-CONTINUOUS

SUMMARY

The generation of food waste grows as the demand for food increases, i.e. it is linked to the entire food production chain. This means that new techniques must be developed to ensure the proper management and treatment of this waste. The anaerobic digestion of food waste is therefore a promising alternative for the treatment of food waste that simultaneously promotes the generation of clean, renewable energy. The aim of this research was therefore to carry out anaerobic digestion tests on food waste from a popular restaurant, checking the potential for energy reuse. BMP tests were carried out in 250 mL reactors to verify the biodegradability of the food waste and in a prototype 408 L anaerobic digester, adapted with agitation and temperature control mechanisms, operating in semi-continuous feeding mode with the food waste. The control variables (solids, pH, VFA, alkalinity, AV/AT ratio and organic load) were collected three times a week over 51 days, totaling 21 samples. COD and SV reductions of 81.63 % and 23.58 % were obtained for the BMP test, respectively, and 82.34 % and 90.22 % for the semi-continuous experiment. It was possible to achieve specific methane production of 0.311 Nm^{3} $_{CH4.kgSVr\wedge}$^{1} for the BMP test and 0.444 Nm^{3} $_{CH4.kgSVr-}$^{1} in the semi-continuous experiment. The daily production of the semi-continuous reactor was, on average, 220 $_{Lbiogas.d\wedge}$^{1} ; and the average volumetric biogas production was 0.540 L.$_{Lr\wedge}$^{1} $_{d\wedge}$^{1} . The stability of the semi-continuous reactor was also verified, with an average AV/AT ratio of 0.49.

Keywords: Biogas; Methane; Food Waste; BMP Test; Prototype Biodigester

1. INTRODUCTION

Brazil's development can take place safely, economically and in compliance with environmental laws, since the country has great energy potential, with emphasis on renewable energy sources such as hydroelectric, wind, solar and biomass.

As a result, the Ten-Year Energy Plan (PDE 2024), drawn up by the Energy Research Company (EPE), included in its main forecasts a greater inclusion and participation of renewable sources in Brazil's energy supply between 2014 and 2024, maintaining, above all, economic growth based on a clean and sustainable energy matrix (EPE, 2015). To this end, Brazilian energy policy has developed numerous incentives for the use of renewable energy sources, focusing mainly on low-carbon technologies and distributed generation (TOMALSQUIN, 2016).

Generally speaking, there are various ways in which the government can encourage renewable energy sources, for example, with fiscal, technological and legal incentives, tax reductions, the dissemination of technology to society, regulations and incentives for industries in the sector (CGEE, 2010). Among the federal incentives that promote the use of clean energies is the Renewable Energy Sources Incentive Program - PROINFA, regulated by Decree No. 5.025 of 2004, which provides resources for new projects based on renewable sources that aim to diversify the Brazilian energy matrix by valuing regional and local characteristics and potential; it also addresses various related concepts, among which it considers, in its Article 5 (BRASIL, 2004):

"PROINFA, instituted with the aim of increasing the share of electricity produced by Autonomous Independent Producers, based on wind power, small hydroelectric plants and biomass, in the National Interconnected System"[...]

PROINFA was a pioneering program that boosted renewable energy sources in the country, especially wind power. It also revealed Brazil's potential for using other alternative sources, such as hydro, solar and biomass, given that Brazil's territory has continental characteristics, thus providing large areas for the most diverse applications. Biomass for energy purposes is among the renewable sources with the greatest possibilities, due to the large quantity and diversity of material available in the country, since biomass comprises all plant matter, animal waste and organic matter contained in industrial and urban waste (MME, 2016).

Currently, the share of energy from biomass in Brazil is 13,425 MW (8.87%) of the total energy generated and it still has great future potential, reaching around 60,000 MW of capacity by 2050 (ANEEL, 2016; TOLMASQUIN, 2016). The main source of biomass comes from the sugar-alcohol and pulp and paper sectors. On the other hand, there is a great diversity of biomass that can contribute to energy generation. For example, a study carried out by the Brazilian Association of Biogas and Biomethane - ABIOGAS reports that the energy use of biogas from the anaerobic digestion of residual biomass could supply approximately 10% of all energy in Brazil (ARAGON, 2015).

Table 1 shows some of the energy projects, taken from the Generation Information Bank (BIG) of the National Electric Energy Agency (ANEEL).

Table 1. Part of the energy projects installed in Brazil.

Source	Type	No. of power plants	Capacity Installed (KW)	Total percentage (%)
	Bagasse Cane	398	10.897.104	6,8627
Biomes	Biogàs - AGRI	3	1.822	0,0011

sa	Black Liqueur	17	2.261.136	1,4240
	Waste Forestry	50	386.525	0,2434
	Biogàs - RA2	11	2.099	0,0013
	Biogàs - RU3	15	114.680	0,0722
Eòlica	Cinèticado wind	413	10.134.742	6,3826
Solar	Radiation Solar	42	23.008	0,0144
Water	Potential Hydraulic	1243	97.368.390	61,320

Source: Adapted from ANEEL (2017). [1]Agroindustrial;[2] Animal Waste;[3] Urban Waste

Table 1 shows that the main source of energy in the electricity matrix is hydropower, which accounts for just over 60% of all the energy generated in the country. As for the participation of biogas, we can see that it is incipient, although it plays an important role in energy generation, with its main contribution coming from landfills and sewage treatment plants. Similarly, it is known that the energy use of agricultural waste allows the integration of food production with a clean, sustainable and economical source of energy, as well as adding value to rural-based production chains, however, enterprises based on Animal Waste Biogas (AR) are not yet practiced by the majority of rural producers (TOLMASQUIN, 2016).

On the other hand, since April 17, 2012, the scenario has tended to evolve, given that ANEEL's Normative Resolution No. 482/12 opened up new possibilities for the injection of electricity from distributed generation into the national interconnected system (SIN), regulating the sale of energy generated by mini and micro-generation, and is considered a regulatory milestone with regard to access by small energy producers to the distribution networks (ANEEL, 2012). In addition, in March 2016 ANEEL's Normative Resolution 682/15 came into force, amending and supplementing Normative Resolution 482/12, resulting in a number of benefits for micro and mini power generation. Among the changes that benefit generators of electricity from biogas and biomethane are the creation of the remote self-consumption and shared generation modalities, the possibility of offsetting credits between headquarters and branches, the formation of consortia and cooperatives between generators, raising the mini generation from 1 to 5 MW and extending the duration of credits from 3 to 5 years (ANEEL, 2015). In this way, investors who are interested in using an anaerobic biodigester to treat waste can obtain income and savings from the energy generated from the biogas produced (ANEEL, 2012; ARAGON, 2015).

Therefore, with the development of regulations for the energy use of biogas and biomethane in Brazil, there has been a parallel drive to improve ways of treating residual biomass, since it is the raw material for the anaerobic digestion process. Considering also that residual biomass must be managed correctly to ensure better environmental quality, since when it is disposed of inappropriately in the environment, it can cause contamination of water resources, greenhouse gas emissions (GHG) and other harm to public health and the environment (YONG *et al.,* 2015).

Currently, the main form of management and treatment for residual biomass is its use in energy conversion processes. Among the technological routes for energy conversion, thermochemical and biochemical are the most widely used, both being characterized as a renewable thermoelectric energy source. The thermochemical route consists mainly of incinerating this waste, but it is still a high-cost technology due to the need to treat the combustion gases; The biochemical route is anaerobic digestion (AD), which consists of microbiological and enzymatic action to mineralize the organic matter, resulting in an organo-mineral biofertilizer and biogas, a mixture of gases rich in methane (CH_4) and highly combustible (WOON & LO, 2016). In addition to being a combustible gas, methane is also considered a Greenhouse Gas (GHG) due to its ability to retain heat in the stratosphere 23 times more than carbon dioxide (CO_2) (IPCC, 2014).

Considering, therefore, that Brazil has a high generation of residual biomass, such as solid urban waste, food waste, agro-industrial waste, agricultural waste and sewage sludge, it is clear that the proper management of this waste must be ensured in order to minimize possible impacts on the environment. In this context, the process of anaerobic digestion as a form of treatment and the energy recovery of biogas are fundamental tools for the sustainable development of the country, since it reduces the contaminant load of waste, allows methane to be oxidized to carbon dioxide and contributes to greater energy security. It should be noted that the country has recently been developing policies to encourage greater use of this technology in the various sectors of the national economy, enabling energy producers to market it and, above all, promoting the consolidation of the biogas and biomethane production chain in the country.

2. OBJECTIVES

The main objective of the research was to evaluate the process of anaerobic digestion of food waste, with the aim of understanding the behaviour of the variables involved in the process, in order to obtain the best efficiency for stabilizing the waste and the highest levels of methane in the biogas.

Specifically, the aim was to:

- To evaluate the physico-chemical and microbiological characteristics of the food waste, namely: pH, total alkalinity (TA), volatile acidity (VA), total and volatile solids (TS and VS), chemical oxygen demand (COD), macro and micronutrients;
- Verify the biodegradability of food waste in bench-top anaerobic reactors by means of *biochemical meth*ane potential (BMP) tests;
- To evaluate the efficiency of a prototype anaerobic biodigester operated in semi-continuous feeding mode with food waste;
- Analyze the parameters that influence the anaerobic digestion process of food waste, such as humidity, temperature, pH, daily organic load, total alkalinity and volatile acidity;
- Monitoring the stability of the anaerobic digestion process in the prototype anaerobic reactor;
- Determine the amount of methane (CH_4), carbon dioxide (CO_2) in the biogas from the batch reactors; and the levels of methane (CH_4), carbon dioxide (CO_2) and hydrogen sulphide (H_2S) present in the biogas from the prototype biodigester;

3. LITERATURE REVIEW

3.1 . NATIONAL SOLID WASTE PANORAMA

Solid waste can be defined based on the text of the National Solid Waste Policy (PNRS), established by Federal Law No. 12.305/2010, regulated by Federal Decree No. 7404/2010, which addresses various related concepts, as well as in its Article 3 (BRASIL, 2010):

[...] XVI - solid waste: discarded material, substance, object or good resulting from human activities in society, the final destination of which is solid or semi-solid, as well as gases contained in containers and liquids whose particularities make it unfeasible to discharge them into the public sewage system or bodies of water, or require technical or economically unfeasible solutions in view of the best available technology.

VII - environmentally appropriate final disposal: waste disposal that includes reuse, recycling, composting, recovery and energy use or other destinations permitted by the competent bodies of Sisnama, SNVS and Suasa, including final disposal, observing specific operational standards in order to avoid damage or risks to public health and safety and to minimize adverse environmental impacts; [...]

Today, solid urban waste is one of the main topics of environmental concern. The National Survey of Basic Sanitation (PNSB), carried out by the Brazilian Institute of Geography and Statistics (IBGE) in 2008, revealed that in Brazil most municipalities (50.8%) dispose of solid urban waste in open dumps; 22.5% dispose of it in controlled landfills, and only 27.7% in sanitary landfills (IBGE, 2010).

The generation of municipal solid waste (MSW) in Brazil in 2014 was approximately 78.6 million tons, an increase of 2.9% compared to 2013. Of this total, more than 51% (by weight) is made up of putrescible organic material (ABRELPE, 2013; ABRELPE, 2014). Naturally, in developing countries, the composition of MSW is mostly organic, with food waste (FF) as one of its main constituents due to the urbanization process (SANTOS *et al.*, 2014; XU et *al.*, 2015). Furthermore, Borges and Guedes (2008) state that the rate of growth of waste generation is greater than the rate at which the population grows.

3.1.1 Food waste

Food waste (FF), also known as human waste, is generated in large quantities, mainly in food establishments such as commercial and collective restaurants, steakhouses, pizzerias, snack bars and bars. Their composition varies according to local eating habits. In Brazil, they generally consist of cereals, meat, pasta, sausages, eggs, fruit and vegetables. These come from the leftovers of food that has been prepared and not consumed and food that has been served and not consumed. This waste is mainly influenced by the lack of planning of the number of meals, the typical nature of the food, the way it is prepared, the absence of quality indicators, purchases made without criteria, among others (BRADACZ, 2003; ZANDONADI & MAURICIO, 2012). In addition, the generation of food waste is inherent to population growth, as it is linked to all stages of the food chain: thus, the greater the demand for food, the greater the generation of food waste (ZHANG *et al.*, 2014).

Tchobanoglous *et al.* (1993) report that due to the organic content of food waste, with between 70 and 90% volatile solids, when it is improperly managed it can cause serious environmental problems, since the products of its degradation are precursors of water contamination, siltation of rivers and lakes, the proliferation of macro and micro vectors, the generation of odors and the emission of Greenhouse Gases (GHG). On the other hand, the biodegradable characteristics and richness in nutrients present in food waste allow it to be recycled and converted into products with greater added value, whether in the form of organic fertilizer or renewable energy (YONG *et al.*, 2015; WOON & LO, 2016).

Food waste can be utilized through different technological routes that promote the recovery of energy and nutrients. Among them, the biological route has become more relevant, mainly through the anaerobic digestion process, since it mineralizes the organic matter and simultaneously produces biogas containing methane (CH_4) that can be used for energy and an effluent rich in nitrogen, phosphorus and other minerals that can be used to improve the fertility and texture of soils (WOON & LO, 2016).

Therefore, anaerobic digestion technology is a fundamental tool for solid waste management in Brazil, since it makes it possible to treat around 50% of waste that would otherwise be destined for landfill and/or disposal in open dumps. It also promotes the mitigation of GHG emissions and, of course, produces renewable energy.

3.2 ANAEROBIC DIGESTION

It consists of a natural microbiological process that provides enzymatic and metabolic interactions on organic compounds (waste biomass), converting them mainly into stabilized matter, methane (CH_4) and carbon dioxide (CO_2). In this process, methane formation occurs in environments where oxygen, nitrate and sulphate are not available as electron acceptors (TCHOBANOGLOUS *et al.*, 1993; CHERNICHARO, 2007; PITK *et al.*, 2013).

Anaerobic digestion represents an ecologically balanced system in which various microorganisms operate symbiotically in two stages (acid digestion and methanogenic digestion), in which at least three physiological groups of microorganisms act: fermentative bacteria (acidogenic), syntrophic bacteria (acetogenic), and methanogenic microorganisms. Each group has its own specific functions, operating in four sequential stages: hydrolysis, acidogenesis, acetogenesis and methanogenesis (TCHOBANOGLOUS *et al.*, 1993).

Figure 1 shows the 4 stages of the digestion process divided into the 2 stages, highlighting the microorganisms involved in the process.

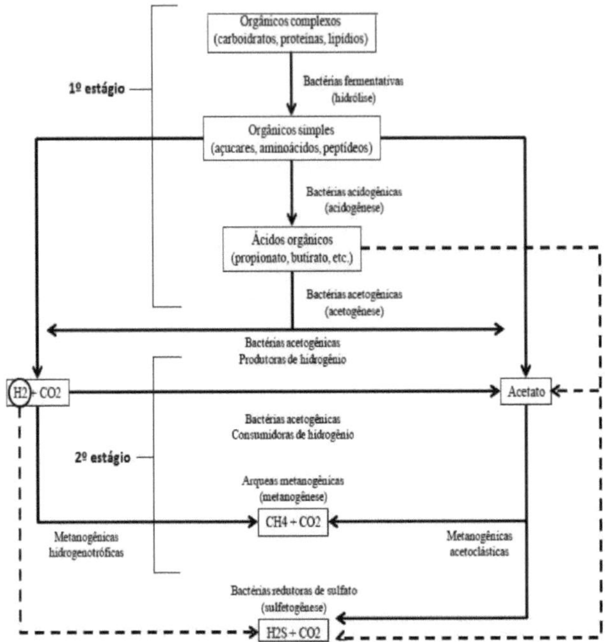

Figure 1: Stages and steps in the anaerobic digestion process.

Source: Adapted from Chernicharo, 2007.

In hydrolysis, fermentative bacteria excrete extracellular enzymes that reduce complex organic polymers into simpler compounds, where proteins, carbohydrates and lipids are broken down into amino acids, sugars and fatty acids, respectively. It is the most active stage of the anaerobic process and is also responsible for the overall speed of the reaction (AMANI et al., 2010). Therefore, the microorganisms involved in acidogenesis are responsible for metabolizing the diversity of products solubilized in hydrolysis, forming carbon dioxide, hydrogen, propionic acid, butyric acid, acetic acid, lactic acid, valeric acid and sulphuric acid. This stage also involves facultative microorganisms, which are essential for oxidizing the organic substrate by the aerobic route and consuming dissolved oxygen, preventing it from becoming a toxic substance in the methanogenic stage (OLIVEIRA, 2012). The microorganisms involved in acidogenesis proliferate very quickly, around 30 to 40 times faster than methanogens, and can survive in extreme conditions such as low pH, high temperatures and high organic loads (AMANI et al., 2010).

In the acetogenesis phase, syntrophic microorganisms convert intermediate compounds (fatty acids and alcohols) into acetate, hydrogen, formate and carbon dioxide, which will be consumed by methanogens. The syntrophic relationship between these microorganisms and the methanogens is fundamental to ensuring better performance for the process (AMANI et al., 2010). Finally, methanogenic microorganisms (strictly anaerobic prokaryotes) convert the acetate and hydrogen produced in the previous stages into methane and carbon dioxide (CHERNICHARO, 2007; DEUBLEIN & STEINHAUSER, 2008). In this process, methanogenic microorganisms play an important role, converting dissolved organic carbon into methane (insoluble gas) and promoting the maintenance of partial hydrogen pressure so that fermentative and acid-forming bacteria can consume it and produce more soluble oxidized products that will be the substrate for methanogens, thus maintaining the balance of the fermentation reaction (CHERNICHARO, 2007; APPELS *et al.*, 2012).

It is worth noting that in the anaerobic digestion process, the sulphate reduction stage (sulphogenesis) can also occur, especially when degradation occurs in protein-rich substrates. The Sulphate Reducing Bacteria (SRB) involved in this stage are able to use the entire chain of volatile fatty acids, hydrogen, methanol, ethanol, glycerol, sugars, amino acids and acetate for their metabolism, making them common competitors for substrate with the acetogens and methanogens, disrupting the symbiotic relationship between these microorganisms, leading to lower overall efficiency in the process and, consequently, lower methane levels (CHERNICHARO, 2007).

Another relevant aspect that can affect the symbiosis between the microorganisms involved is the degradation rates, which must be the same in both stages, shown in Figure 1, for the process to be efficient. Otherwise, if the process goes too fast in the first stage, the concentration of acids and carbon dioxide rises and the pH drops below 7.0, so acid fermentation also occurs in the second stage. Similarly, if the second stage occurs too quickly, showing that too many microorganisms from the first stage are present in the second stage, methane production decreases and new microorganisms from the second stage need to be introduced to restore the balance (DEUBLEIN & STEINHAUSER, 2008).

The parameter that is often used to verify ecological equilibrium in anaerobic systems is the concentration of volatile fatty acids (VFA). VFA are intermediate products of the anaerobic digestion process, originating from the breakdown of carbohydrates, proteins and lipids and are characterized by their low molecular weight, such as propionate, butyrate and other compounds lower than acetate, such as short-chain acids (AQUINO and CHERNICHARO, 2005). When the system is balanced, with a sufficient population of methanogenic bacteria and favorable conditions, the VFA are consumed as soon as they are formed, so they don't accumulate in the system and the pH remains neutral. On the other hand, when the system is under conditions of stress and kinetic limitations of the methanogenic microorganisms, the VFA are generated at a higher rate than they are consumed, accumulating in the medium and causing a drop in pH, which sours the reactor and the activity of the methanogenic microorganisms ceases (CHERNICHARO, 2007).

3.2.1 Thermodynamic aspects

Methanogenic microorganisms use a limited group of substrates for their growth and methane formation, mainly acetate (Eq. 1), hydrogen and carbon dioxide (Eq. 2) and to a lesser extent methanol (Eq. 3), formate (Eq. 4), methylamines (Eq. 5) and carbon monoxide (CO). The main mechanisms involved in the formation of methane are shown in the equations below:

$$CH_3COOH \rightarrow CH_4 + CO_2 \quad (1)$$

$$4H_2 + CO_2 \rightarrow CH_4 + 2H_2O \quad (2)$$

$$4CH_3OH \rightarrow 3CH_4 + CO_2 + 2H_2O \quad (3)$$

$$4HCOOH \rightarrow CH_4 + 3CO_2 + 2H_2O \quad (4)$$

$$4(CH_3)3N + H_2O \rightarrow 9CH_4 + 3CO_2 + 6H_2O + 4NH_3 \quad (5)$$

Eq. 1 refers to acetate shear, which is responsible for 70% of the methane produced by acetoclastic methanogens. In the absence of hydrogen, the methyl group is reduced to methane and the carboxylic group is oxidized to carbon dioxide. Eq. 2 represents the reduction of carbon dioxide when hydrogen is available. co_2 acts as an acceptor of hydrogen atoms that have been removed from organic compounds, forming methane and water (TCHOBANOGLOUS *et al.*, 2003; CHERNICHARO, 2007). These are the main pathways responsible for methane formation.

According to Van Haandel & Lettinga (1994), the stoichiometry of biological reactions is more advantageous in anaerobic treatment than in aerobic treatment. In anaerobic treatment there is no need to introduce an oxidizer. The process produces less sludge and methane, which can be used as fuel. However, the intrinsically favorable stoichiometry of anaerobic digestion alone does not make it a suitable alternative for removing organic material. Basically, there are two other important factors: (1) the organic material removal efficiency must be high, so that there is a low concentration of residual organic material in the effluent from the treatment system, and (2) the removal rate must be high, so that it can be achieved in a reactor with a short residence time, i.e. a reactor with a small volume. Both factors are linked to the kinetics of organic material removal and the operational and environmental conditions in the sewage treatment system.

3.2.2 Natural and operational aspects

In general, the overall efficiency of the anaerobic digestion process and the methane content in the biogas depend fundamentally on the environmental and operating conditions in the anaerobic reactors. Therefore, the quality of the biogas in practice is associated with various factors, such as: substrate composition, particle size, organic load, pH, alkalinity, humidity, macro and micronutrient balance, temperature, hydraulic retention time and agitation frequency. Similarly, substrates that are difficult to degrade biologically, especially lignocellulosic materials, find hydrolysis to be the limiting step in the process, since hydrolytic enzymes are not efficient at breaking down these materials (VAN HAANDEL & LETINGA, 1994; DEUBLEIN & STEINHAUSER, 2008).

Table 2 shows the main parameters involved in the anaerobic digestion process.

Table 2. Parameters involved in the anaerobic digestion process.

Parameters	Value ideal	Unit	Observations
Temperature - Mesophilic - Thermophilic	30 - 40 40 - 70	°C	The upper limit of mesophilia is the ideal temperature for optimum biogas production.
pH	6,3 - 7,9	-	Outside these levels methanogenic microorganisms do not develop
Alkalinity	1000 a 5000	mg CaCO3/L	Neutralizes residue acidity variations
Volatile Acidity	500 a	mg	Higher concentrations will inhibit

	2000	CH3COOH/ L	acetate and biogas production.
Acidity/Alkalinity (AV/AT)	0,1 a 0,5	-	Represents stability for anaerobic processes, higher values represent an accumulation of acids in the reactors.
C/N	20 a 30	-	Higher ratios lead to the consumption of N by methanogens , reducing methane production.
Organic load - Mesophilic - Thermophilic	 0,4 a 6,4 1,0 a 7,5	kgSV/m r^{3} ^{1}.d	Microorganisms are inhibited if the organic load is too high.
HRT	9 a 95	days	It varies depending on the substrate, temperature and the type of digestion system.

Source: Adapted from Pecora, 2006; Amani et al., 2010; Perovano and Formigoni,

2011; Cabbai *et al.*, 2013.

3.2.2.1 Temperature

Temperature is one of the most significant parameters influencing the anaerobic digestion process, not only because it restricts enzymatic and coenzyme activity, but also because it influences methane production and digestate quality (APPELS et al., 2011).

Bacteria can grow at minimum, optimum and maximum temperatures. At optimum temperatures, the enzymes are in their most active form; at minimum temperatures, the enzymatic efficiency in the conversion rate of organic matter is significantly reduced, limiting the process in general; and at maximum temperatures, protein denaturing (destruction of the molecular arrangement) can occur, leading to cell death (ALVES, 2008).

Temperature has a direct influence on the hydrolysis stage, where if there is a reduction in enzyme activity, the overall reaction speed of the anaerobic degradation process may be limited, given that hydrolysis is the initial stage and is responsible for making the substrate available for the other stages (CHERNICHARO, 2007). For example, the reduction in COD fell from 88.5 to 68.1% when the temperature was reduced from 24 to 16 °C in a static granular bed reactor for the stabilization of swine manure (LIM & FOX, 2011).

Similarly, methanogenic *archaea* are very sensitive to temperature variations, and develop mainly at mesophilic (30 - 40 °C) and thermophilic (40 - 70 °C) temperatures. The thermophilic arrangement has higher rates of conversion of solids into biogas, however, the effluent is of poor quality and the system is more susceptible to instabilities, which can inhibit biogas production. The mesophilic process, on the other hand, despite having better stability and greater microbial diversity, can have low methane levels due to the greater demand for nutrients, due to the wide variety of microorganisms operating (MAO *et al.*, 2015). For reactors operated below 30 °C (psychrophilic temperatures), each degree centigrade reduction in temperature implies an 11% drop in the maximum rate of anaerobic metabolism, affecting the fraction of organic solids that can be metabolized, the gas transfer rate and the sedimentation characteristics of the biological sludge (VAN HAANDEL & LETTINGA, 1994; TCHOBANOGLOUS *et al.,* 2003). Silva (2012) and Bouallagui et al. (2004) report that, on average, optimal microbial growth is around 35 to 37°C, promoting the maximum rate of biogas production.

3.2.2.2 Hydrogen Potential, Alkalinity and Volatile Fatty Acids

The pH directly affects the level of metabolic activity of methanogenic microorganisms, which can cease at values outside the range of 6.3 to 7.8. Even so, in the fermentation phase where acid digestion takes place, the microbial population tolerates pH at more acidic or alkaline levels. However, a pH below 4.5 stops the activity of all the microorganisms involved in the process. The pH influences not only the production of biogas but also its quality, where values below 6 result in a biogas poor in methane (LEMA & MÉNDEZ, 1997). Acids, if increased beyond the digestive capacity of the microbial population specific for their degradation, put the stability of the fermentation at risk. The main cause is the imbalance between the production of acids and bicarbonates in the first stage of digestion, where the alkalinity provided by the bicarbonates is not enough to neutralize the acids, causing a drop in pH, where acidic fermentation prevails (PAES, 2003). This is due to the logarithmic scale of pH, where small decreases in pH involve a high consumption of alkalinity, reducing the buffer solution of the medium. Likewise, if the medium is alkaline (pH > 8), the efficiency of the system decreases, since it can influence the production of ammonia, a toxic component for the anaerobic process at concentrations above 150 $mg.L^{-1}$ (CHERNICHARO, 2007).

Volatile fatty acids (VFA) and alkalinity are important indicators of stability in anaerobic reactors. Volatile acidity, quantified in mg of acetic acid per L, indicates the concentration of acids and measures the capacity of the anaerobic fermentation process to resist pH increases when a base is added. Total alkalinity, quantified in mg of calcium carbonate per L, indicates the concentration of alkalis involved in fermentation and measures the system's ability to resist lowering the pH when acids are added. These two indicators are the most important for monitoring anaerobic processes (AMANI et al., 2010).

Therefore, it is necessary to dose the amount of substrate to be added to the reactor, in which the estimated reference is the ratio of volatile acids x alkalinity (VA/AT) existing in the fermentation, whose value, according to Sànchez *et al.* (2005), should be between 0.1 and 0.5 so that the system maintains a balance in the reactions of production and consumption of the compounds.

3.2.2.3 Biomass Characteristics and Organic Load

The content of rapidly biodegradable matter, such as carbohydrates, proteins and lipids, is related to the development of microorganisms and therefore qualitatively and quantitatively affects the production of biogas (MACIEL and JUCA, 2011).

According to Tchonobaglous *et al.* (1993), food waste falls into the category of rapidly biodegradable waste. Zhang *et al.* (2014) mention that the volatile solids (VS) concentration of a given substrate also refers to its biodegradable organic matter content. Other authors use the Total Organic Carbon (TOC) parameter to assess the organic matter content of substrates (FONSECA *et al.*, 2006; GRIGATTI *et al.*, 2004). Even so, the most widely used parameter for this purpose is the characterization of solids, mainly due to the low cost and less onerous nature of the test (HAMILTON, 2012).

Table 3 shows the data obtained by different authors on the characterization of the solids content of food waste.

Table 3. Solids content of food waste.

Parameters	**Zhang et al., 2007**	**Li et al., 2010**	**Haider et al., 2015**
ST (% b.s.)	30,90 +- 0,07	24,00	24,00
SV2 (% b.s.)	26,35 +- 0,14	23,20	22,13
STV3 (%)	85,30 +- 0,65	94,10	92,20

[1] Total Solids;[2] Volatile Solids;[3] Total Volatile Solids; b.s: dry basis.

Angelidaki *et al.* (2009) report that materials with STV contents above 80% have excellent biodegradability profiles and can be used in anaerobic systems. Fernàndez, Pérez and Romero (2008) show that reactors operating with a total solids (TSS) content of between 20 and 30% achieve more effective biogas production and are richer in methane, due to the presence of moisture which allows the active microorganisms to absorb more nutrients. In addition, humidity is essential to ensure the efficiency of metabolic processes, mobilization and microbial growth (SILVA, 2012).

The organic load is represented by the amount of volatile solids added to the reactor. As the organic load increases, biogas production tends to increase, respecting the maximum limit of up to 6.4 $kg.m^{-3}.d^{-1}$, otherwise, without taking due care, the balance of the digestion process can be seriously affected with microbial inhibition. Sudden changes in the type of substrate to be added to the reactor can also lead to inhibition of microbial activity in the fermentation phases (MAO *et al.*, 2015). Therefore, it is essential that the reactor's sizing is appropriate to the input parameters, such as the type of substrate and the daily organic feed load.

3.2.2.4 Nutrients and C/N Ratio

The presence of macro and micronutrients is fundamental for bacterial metabolism, growth and activity. The main nutrients for the life of anaerobic microorganisms are carbon, nitrogen and phosphorus, and a series of mineral elements such as sulphur, iron, potassium, sodium, calcium and magnesium, provided mainly by the hydrolysis of carbohydrates, proteins and lipids (PARK, 2012). These nutrients, at balanced levels, positively influence the development of microorganisms and, consequently, the process of digestion and biogas production.

The C/N ratio is a significant indicator of the digestion capacity and potential yield of biomass, where its value varies depending on the biomass. An adequate ratio for the development of microorganisms is in the range of 20 to

30 (VERNA, 2002).

Carbon is the main source of food for microorganisms and the main component of biogas. It is mainly derived from carbohydrates contained in the biomass, while nitrogen is responsible for synthesizing the organisms' proteins. Under conditions of excess nitrogen (low C/N ratio), nitrogen accumulates in the medium, usually in the form of NH_3, which inhibits the growth of methanogens. On the other hand, in conditions of limited nitrogen availability (high C/N ratio), the microorganisms are unable to metabolize the carbon present, which leads to inefficiency in the process (SGORLON *et al.*, 2011). In general, the absence of nutrients in the mixture reduces the growth of methanogenic microorganisms, leading to an intense accumulation of organic acids in anaerobic reactors, thus reducing the quality of the biogas.

Food waste is very biodegradable, but it lacks the nutrients and mineral salts that are essential for the development of microorganisms. Inhibition can occur when only food waste is biodigested for long periods of operation. The reasons for the inhibition are the imbalance of nutrients in the fermentation, i.e. the trace elements (Zn, Fe, Mo, etc.) are insufficient, the macronutrients (Na, K, etc.) are excessive (ZHANG *et al.*, 2011; ZHANG *et al.*, 2013; EL-MASHAD & ZHANG, 2010). Because of this, several authors indicate the use of two or more substrates in the DA process to take advantage of the synergy of the mixtures, compensating for the nutritional deficiencies of the substrates and supporting the growth of the microorganisms, consequently improving the quality of the biogas (ZHANG *et al.*, 2014; RATANATAMSKUL *et al.*, 2015; CHEN et *al.*, 2016).

Currently, most of the studies found in the literature on the DA of food waste report on codigestion with different substrates, for example: food waste with cattle manure (BOULLAGUI *et al.*, 2004; EL-MASHAD & ZHANG, 2010), food waste with rice husks (HAIDER *et al.*, 2015) and food waste with straw (YONG *et al.*, 2015). In this way it is possible to balance the nutrients in the medium, providing higher rates of biogas production.

3.2.2.5 Stirring the biomass

Stirring the biomass in an anaerobic digester has a significant effect, as it promotes the homogenization of the substrate and increases the kinetic speed of anaerobic digestion, accelerating the process of biological conversion. This is due to the uniform heating of the substrate inside the digester as well as the greater ease of material transfer, since the substrate molecules must be absorbed by the surface of the microorganisms and the intermediate and final products must be transported (TCHOBANOGLOUS *et al.*, 2003). In addition, agitation prevents

short circuits (dead zones) occur inside the biodigester, preventing part of the substrate from flowing out without coming into contact with the microorganisms.

It is recommended that only reactors with a volume of less than 50 m^3 operate without agitators; for larger reactors, it is essential to have agitation during the digestion process (DEUBLEIN & STEINHAUSER, 2008). Agitation can be carried out in various ways, by mechanical devices, recirculation of the biodigester contents or recirculation of the biogas itself (BARAZA *et al.*, 2003). It is worth noting that the intensity and duration of agitation are crucial to the procedure. Even so, the

Information on the intensity and duration of agitation on the performance of biodigesters in the literature is quite contradictory, bringing the need for extensive research in this regard (KARIM *et al.,* 2005).

3.2.2.6 Toxic substances and inhibition

There are a wide variety of toxic substances that are responsible for the failure of anaerobic systems. The accumulation of organic acids produced in the first stage of the process is one of the main causes of instability in the anaerobic digestion process, and sudden variations in environmental factors can also interfere in the process, such as pH, temperature, nutrients, organic load and the presence of inhibitory agents in anaerobic reactors, which include high concentrations of ammonia, sulphates and heavy metals, mainly (CHEN *et al.*, 2008).

In general, toxic compounds can have different effects on the biota of anaerobic systems. They can have a bactericidal effect when the bacteria cannot withstand the presence of the contaminant, or a bacteriostatic effect when the bacteria are able to adapt under certain concentrations of the toxic contaminant and therefore resist and recover to normal activity after a certain resilience time (AMANI et al., 2010).

Above all, the best way to combat toxic agents is to use an antagonistic agent, where one toxic product is annulled in the presence of another. For example, sodium (Na) and potassium (K) in a mixture cancel each other out, reducing the toxic effect of both. Among the main inhibitory agents are nitrates, cyanides, phenols, sodium, potassium, calcium, magnesium, ammoniacal nitrogen, oxygen and heavy metals only when soluble.

However, the toxicity of these components varies depending on their concentration in the environment (FORESTI, 1993).

3.2.2.7 Starting up anaerobic reactors

The start-up of anaerobic or split reactors after maintenance is generally the most critical stage for the success of the process, since it is the stage responsible for the start of all microbial activity. This stage depends on several factors, such as the source of microorganisms (inoculum) to be used and the initial operating mode, for example (LOPES et al., 2004). It is essential that the inoculum has a community of facultative and methanogenic microorganisms to guarantee the initial performance of the anaerobic digestion process (AMANI et al., 2010). It is considered that the start-up of anaerobic systems is an area that still requires intensive research. Amani et al. (2010) proposes, however, that it should be evaluated in conjunction with other criteria, such as different forms of biomass pre-treatment, in order to technically investigate the real performance of anaerobic reactor start-ups.

Silva (2014) investigated the effects of using anaerobic sludge and cattle waste to start up a reactor to treat food waste. He found that anaerobic sludge from a UASB reactor at a sewage treatment plant had the highest rate of acclimatization of the inoculum to the substrate, and that methane production increased in the first few days. He also points out that, for best results, using an inoculum of the same nature as the substrate allows the acclimatization phase to be quicker, as well as providing better conditions for microbial development. However, there is still a lot of controversy over which is the best inoculum for acclimatizing a given substrate, as well as the optimum ratio between them.

3.2.2.8 Anaerobic digestion of food waste

The anaerobic digestion of food waste is a complex process that must simultaneously digest carbohydrates, proteins and fats in a single-stage system. The process is closely influenced by several key parameters, such as temperature, pH, concentration of VFA, ammonia and nutrients, among others. It is extremely important to maintain the key parameters at appropriate levels for a long operating time (ZHANG et al., 2014). .

3.3 BIOGAS

Biogas has combustible characteristics due to the presence of methane gas (CH_4) in its composition, which has a calorific value of approximately 35,800 kJ/m^3 . However, methane is considered a precursor of the greenhouse effect because it retains heat in the stratosphere 25 times more than carbon dioxide. Therefore, energy recovery from biogas is a fundamental tool for sustainable development, since it allows methane to be oxidized into carbon dioxide, taking advantage of its calorific value and contributing to greater energy security (IPCC, 2014; MACHADO, 2011). In addition to methane, biogas is also composed of other gases, such as carbon dioxide (CO_2), hydrogen sulphide (H_2S), ammonia (NH_3) and other gases in trace concentrations. Table 4 shows the main constituents of biogas and their respective characteristics.

Table 4. Properties of the constituent elements of biogas in combustion.

Component	Content	Observations
CH_4	50 - 75 %	High calorific value;
CO_2	25 - 50 %	Low calorific value; Modifies properties anti-detonation when used in engines; Causes corrosion in the presence of water vapor;
H2S	0 - 0,5 %	Corrosive corrosive effect on structures metal; Emission ofSO_2 and H2S in incomplete combustion;
NH_3	0 - 0,05 %	NOx emissions during combustion; Increases the properties anti-detonation;
Water vapor	1 - 5 %	Causes corrosion in equipment;
N_2	0 - 5 %	Decreased calorific value; Increases properties properties anti-detonation;
Siloxanes	0 - 50mg/m^3	It acts as an abrasive and damages motors;

Source: Deublein and Steinhauser (2008).

Biogas can be used in any activity to generate heat, motive power and electricity (VERNA, 2002; PECORA, 2006). However, for many applications of biogas, refinement is recommended, mainly to remove CO_2, water and H_2S. It should be mentioned that in some cases biogas can be used *in its natural* form, such as for direct firing in cogeneration boilers, where the removal of contaminants is less necessary. On the other hand, for more noble uses, such as in fuel cells, the quality of the biogas must meet more stringent quality standards (ZANETTE, 2009). Nevertheless, it is interesting to remove H_2S and water, as they drastically reduce the useful life of the components used, such as pipes, pumps, motors, etc.

Souza & Schneider (2016) report that levels of H_2S above 250 ppm in biogas make it unfeasible to use it in conversion engines, due to the intense abrasion on metal parts and the deterioration of lubricating oils. In addition, after the combustion of biogas, sulphur oxides (SO_2 and SO_3) are released, which are even more toxic than H_2S (WEILAND, 2010). Similarly, the importance of reducing the moisture content of biogas is due to the consequent condensation under high pressure which promotes wear and tear on parts, as well as its reaction with sulphur oxides resulting in sulphurous acid (H_2SO_3), which has very high corrosive power and is a precursor to acid rain. As for the need to remove CO_2, this is basically to increase the energy value of the biogas, since its presence has a diluting effect on the calorific value of the biogas (IEA - BIOENERGY, 2001).

In Brazil, biogas is considered raw gas, according to ANP Resolution No. 8/2015, and therefore needs to be refined to meet certain requirements for commercial fuel, known as biomethane (BRASIL, 2015). Ryckesbosch *et al.* (2011) report that the treatment of biogas aims to increase its commercial value and, above all, its energy value.

Table 5 shows the biomethane specifications according to ANP requirements.

Table 5 Biomethane specifications according to ANP Resolution No. 8/2015.

Features	Unit	Limits	
		Northern Region	Other regions
Methane	% v/v	90.0 to 94.0 min	96.5 min
Oxygen	% v/v	0,8	0,5
CO_2	% v/v	3,0	3,0
Sulphydric gas	mg/m^3	10	10
Dew point	°C	-45	-45

Source: Adapted from BRASIL (2015).

As Table 5 shows, there is a need to refine biogas so that it meets the specific requirements of biomethane fuel. To this end, there are various technologies capable of refining biogas, removing its impurities and increasing its calorific value.

3.3.1 Stages of the biogas purification process

Currently, there are several technologies available for refining biogas, removing the non-combustible components and consequently increasing its calorific value and commercial value. According to Ryckesbosch et al. (2011), the purification process essentially consists of two main stages:

i) Removal of non-combustible gases;

ii) Methane enrichment.

Basically, the stages of the process can be summarized as desulphurization (removal of sulphuric acid), drying (water drainage) and CO_2 removal. According to Souza & Schneider (2016), the characteristics of each stage are:

- Desulphurization is considered to be the fundamental stage for the use of biogas and can be carried out by different mechanisms, such as biodesulphurization, external chemical washing of the biogas, internal chemical desulphurization and desulphurization with activated carbon, the latter being the most commercially used. The aim of this stage is to protect all the equipment involved in the process from the corrosive action of the sulphuric acid present in the biogas.

- The drying of biogas is necessary to avoid the occurrence of water condensation during the process, thus ensuring greater safety and efficiency in energy use. Drying techniques can be carried out by condensation, adsorption or absorption, all of which have advantages and disadvantages depending on the refining requirements.

- CO_2 is removed so that the biogas reaches a higher degree of purity (methane content above 96.5%, known as biomethane), allowing it to be injected into natural gas networks or used in vehicles. CO_2 has zero calorific value and therefore its presence in biogas is detrimental to energy performance. CO_2 removal involves more advanced technologies and requires greater investment than the other stages, so the choice of a particular technology must take into account certain factors, such as methane losses, energy costs, availability and price of inputs.

In this way, the process of purifying biogas opens up new possibilities for its application, since by meeting biomethane specifications, it can be injected into natural gas (NG) networks, a technique that has already been widely used in several countries. Therefore, even if purification adds costs to biogas production, it is still feasible, since the use of biogas reduces NG imports, as well as numerous other environmental, economic and social advantages. From a strategic point of view, it provides energy in a decentralized way and close to the generation points, is a renewable and low-cost fuel, and drastically reduces emissions of polluting gases (SCHUCH, 2012).

3.3.2 Biogas applications

Biogas technology is already being widely used in several countries. In

Europe, the installed capacity of biogas plants exceeds 2000 MW, concentrated mainly in Germany, Sweden and the United Kingdom, which are the most advanced countries in terms of biogas utilization (REF). In Brazil, the use of biogas is still in its infancy, with a capacity of just over 86 MW, mainly from MSW biogas (ANEEL, 2016).

In the western region of the state of Paranà, they have developed a modern and diversified biogas exploitation system, which integrates the biodigesters of rural producers into low-pressure gas pipeline lines, directing the biogas generated to a Mini-Thermoelectric Plant (MCT). This plant purifies the biogas and converts it into electricity and heat for drying grain. The project receives biogas from 33 rural producers, totaling approximately 821.8 m^3 /day. In addition, the methane gas not emitted into the atmosphere can be converted annually into tons of CO_2 equivalent and sold through clean development mechanism (CDM) projects, thus increasing the system's economic viability (SCHUCH, 2012).

Itaipu Binacional (IB), together with the International Center for Renewable Energies (CIBiogàs - ER), has been investing in the development and expansion of biogas technology in Brazil. Recently, they developed a partnership consisting of the installation of the Biomethane Demonstration Unit with the aim of treating the residual biomass from grass pruning, restaurant waste and part of the sanitary effluent generated by the IB facilities. As well as treating the waste, they will be producing biogas that will be converted into biomethane to serve the car fleet and generate thermal energy.

3.3.3 Biogas production estimates

In general, daily biogas production depends on the amount of volatile solids in the biodigester's feed load, since VS represent part of the total solids susceptible to being biodegraded and converted into biogas. Therefore, the most widely used method for estimating biogas generation is based on the volatile solids content, which allows the feasibility of producing biogas from a given substrate to be verified.

Table 6 shows the methanogenic potential for different substrates.

Table 6. Methanogenic potential of different substrates.

Material	Potential Methanogenic Source (m^3 CH4.kgSV^{-1} b.s)	
Wheat straw	0.522Hashimoto	, 1986
Fruits	0,180 a 0,732	
Vegetables	Gunaselaan, 2004 0.190 to 0.400	
Organic fraction	of	
MSW	0,489 Mata-alvarez 2002	
Food waste	0,525	Lissens *et al.*, 2004
Legumes	0,211	Raposo *et al.*, 2006
Food waste	0,363	Neves *et al.*, 2006
Food waste	0,479	Zhang et al., 2011
Food waste	0,410	Zhang *et al.*, 2013
Fish waste	0,441 a 0,482	
Brewery waste	0,316	Kafle *et al.*, 2013
Bread waste	0,306	
Raw swine manure	0,568	Amaral *et al.*, 2016
Cattle manure	0,204	
dairy	0,155	Kafle and Chen, 2016

Horse manure	0,159
Goat dung	
Chicken manure	0,259
Pig manure	0,323

Table 6 shows that food waste has a good potential for biogas production, with specific methane production values higher than most substrates. This estimate can be made through experimental studies, carried out in bench reactors (laboratory scale), such as the *Biochemical Methane Potential* (BMP) test, which aims to biodegrade the substrate and in parallel measure the specific methane production per unit of organic load (COD or Volatile Solids, mainly).

Estimates can also be based on the composition of the substrate. The IEA - Bioenergy (2009) reports that, on average, 72.5% of proteins and 70.5% of fats are transformed into CH_4. Carbohydrates, in turn, are converted into a significant volume of biogas with a methane content of 52.5%. Even so, it should be noted that the composition and quantity of biogas depends fundamentally on the nature of the substrate, the model of the biodigester and the operating conditions.

3.3.4 BMP tests

The BMP Test (Biochemical Methane Potential), although not yet internationally standardized, is a reference analytical method when it comes to obtaining more details about the transformation of organic materials into methane (MACIEL, 2009). The anaerobic digestion process is carried out under optimum degradation conditions and can be considered an accelerated anaerobic digestion process (HENRIQUES, 2004).

The procedure basically consists of monitoring substrate biodegradability, sludge activation, daily biogas production and possible inhibitions during the anaerobic digestion process (ANGELIDAKI et al., 2009). The first methods were based on measuring methane by displacing a liquid, or by displacing plungers in glass syringes, or by measuring pressure using manometers attached to the reactors (ALVES, 2008). The process is usually carried out in bench reactors fed with nutritional supplements, nitrogen, sodium and phosphorus compounds, macro and micronutrient solutions, inoculum and a biodegradable solution (substrate). The internal environment of the reactors is maintained in strict anaerobiosis by removing atmospheric oxygen through the injection of N2 or another inert gas into the reactor flasks (OWEN *et al.*, 1979), 1979). The reactors then remain

incubated at a constant temperature for a minimum hydraulic retention time (HRT) of 30 days for simple substrates and 120 days for lignocellulosic or recalcitrant substrates, although the ASTM E2170- 01 standard recommends a HRT of 51 days or until gas production becomes stable (MELO, 2010). In the BMP tests carried out by Gunaseelan (2004), fruit and vegetable waste was used as a substrate and methane generation curves were obtained which represent more than 90% of the total methane generated over a period of 30 to 40 days. In other words, the TRH can vary depending on the characteristics of each experiment.

In general, the test can be carried out at different temperatures: 10 to 20°C (psychrophilic), 20 to 37°C (mesophilic) and 37 to 70°C (thermophilic). The inoculum:substrate ratio to be used depends on the objectives of the research, and is calculated from the volatile solids content of the samples, such as 1:1 or 1:3, for example. It is worth noting that it is essential to carry out this type of experiment in triplicate due to the great variability of the results when studying systems with living organisms. However, variations in the configurations of the BMP experiment often generate different results, which prevent more consistent comparisons (ANGELIDAKI et al., 2009).

However, the main objective of the test is to obtain the specific production of methane, usually by measuring the volume of methane that can be obtained from a quantity (by weight) of reduced volatile solids (SVr) in the substrate (HAMILTON, 2012).

4. MATERIALS AND METHODS

The methodological approach of this project consisted of intensive theoretical, exploratory and descriptive research, in order to contextualize the current situation of food waste management in Brazil and worldwide, presenting anaerobic digestion technology as an alternative for treatment and energy generation. Participation was made in seminars, congresses and events related to the topic, and the work was carried out at the extension offices of the State University of Western Paranà - UNIOESTE, Cascavel *campus.*

4.1 SUBSTRATE AND INOCULUM CHARACTERIZATION

The food waste was supplied by a popular restaurant in the municipality of Cascavel (PR), shown in Figure 2, located at 24°57'31.6"S 53°27'02.8"W, which serves an average of 700 meals a day, consisting mainly of cereals, pasta, fruit, vegetables, sausages and meat.

Figure 2 - Location of the municipality of Cascavel - Paranâ.

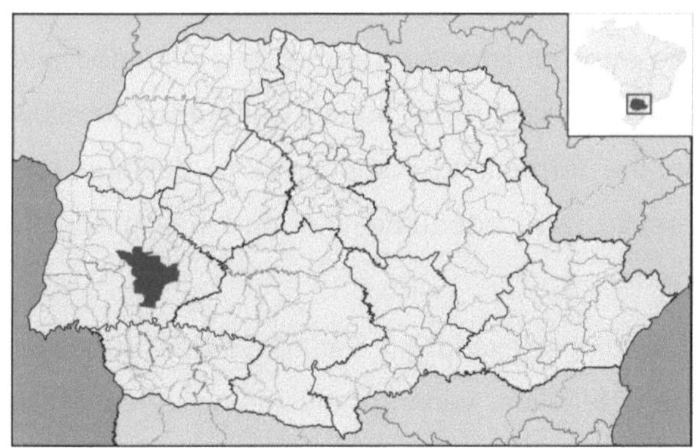

Source: IBGE, 2016.

The collection routine was weekly, taking place every Wednesday for the duration of the experiment. Once collected, the food waste was segregated (bones and lumps removed), processed in a blender operated at speeds of 2000 and 22000 RPM to reduce particle size, packed in 2-liter PET bottles and kept in a refrigerator at -18°C in order to preserve its characteristics for feeding on days when the substrate was not collected. Feeding took place from Monday to Friday, days on which the restaurant serves and therefore generates waste. The feeding frequency was adopted according to the waste generated by the restaurant.

The inoculum used was supplied by a cooperative that has an anaerobic biodigester for the treatment of pig manure.

4.2 BIODIGESTER START-UP AND FEEDING FREQUENCY

The start of the anaerobic process in the prototype biodigester was carried out with the help of the effluent from an anaerobic biodigester in operation, fed with swine manure from a cooperative in the region. Inoculation took place in a ratio of 1:210:195 (v/v) in substrate (food waste), inoculum and water, respectively. Therefore, the percentage of inoculum used in the mixture was above the recommendation contained in the methodology presented by Foster-Carneiro, Perez and Romero (2008). Based on this, a low organic load was set and the pH and AV/AT ratio were monitored to guide the level of the load to be applied. Based on the results obtained, the frequency and intensity of the organic load to be added was varied to allow the microorganisms present in the fermenting compounds to develop and adapt. If the pH and AV/AT results were adequate, the load was gradually increased. If not, the load was reduced.

The biodigester received daily feedings (except on Saturdays and Sundays) of substrate with the same characteristics, varying only the volumes which were gradually increased until the process stabilized, when it was possible to obtain the highest levels of methane. The indicators for the daily increase in organic load and process stability were pH, the ratio of volatile acidity to total alkalinity (VA/TA) and biogas production. After stabilizing the anaerobic digestion process in the biodigester, the 51-day analysis stage began.

For the feeding process, the food waste was first thawed on the respective feeding days, in the quantities needed to feed the biodigester and collect samples for the input physico-chemical analysis. After the biodigester was fed, a quantity close to the inserted substrate was purged from the outlet duct (Figure 2), from which a sample was collected and stored to be submitted to the physicochemical characterization analyses of the outlet or submitted to the tests pertinent to the stability of the biodigester.

4.3 PHYSICAL, CHEMICAL AND MICROBIOLOGICAL ANALYSIS

Inlet and outlet samples were collected three times a week, on Mondays, Wednesdays and Fridays, for a period of 51 days after stabilization of the biodigester under adequate load, for a total of 21 samples.

The samples were stored in a refrigerator at -18°C to preserve their characteristics until the tests were carried out. The analyses of the monitoring variables for the increase in load, stability and solid fractions were carried out at the Sanitation Laboratory at UNIOESTE -

Campus Cascavel and all the analyses were carried out according to the standards established in APHA (2005), described in Table 7.

Table 7. Methodology for analyzing biogas parameters and composition.

Variable	Unit	Method
T	°C	Digital thermometer
pH	-	Potentiometric [1]
$_{Total}$ COD	mg $_{O2}$ L^{-1}	Colorimetric Flow closed [1]
SV	mg L^{-1}	Gravimetric [1]
AV	mg $CH3COOHL^{-1}$	Titrimetric[1]

TOC	mg L^{-1}	TOC =SV/1.8 (base dry)[3]
NTK	mg L^{-1}	Distillation[1]
AT	mg CaCO3 L^{-1}	Titrimetric[1]
Biogas composition	CH4 and co2 (%)	Gas chromatography [2]

Source:[1] APHA (2005); [2] Vaz *et al.* (2003); [3] Cornell Composting (1996).

4.4 ANAEROBIC BIODIGESTER PROTOTYPE

The prototype anaerobic biodigester with a volume of 408 L was designed and assembled by the company BioKohler® Biodigestores, located in Marechal Càndido Rondon, PR. The manufacturer provided the prototype with a digestion chamber, an internal heating coil, a vinyl flask for storing biogas, a mechanical agitator with a timer and an electric drive panel.

At UNIOESTE, the biodigester was complemented with a boiler to heat the recirculating water through the biodigester's internal coil to maintain the temperature, and a U-tube pressure gauge was also installed to monitor the biodigester's internal pressure. The water in the boiler was heated using three 200W electric heaters, in order to keep the internal temperature of the biodigester constant at 29 ± 0.5 °C and prevent fluctuations in the ambient temperature from interfering with biogas production.

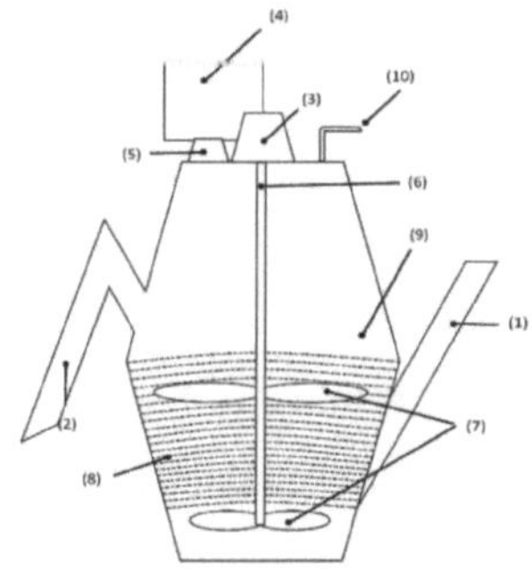

1: Entry
2: Salda
3: Electric motor
4: Control Panel
5: Temperature sensor
6: Rotation axis
7: Agitation step
8: Internal Heating Coils
9: Digestion chamber
10: Biogas outlet

Figure 3 - Photo and diagram of the anaerobic biodigester prototype.

Source: Author.

As shown in Figure 3, the vertical rotation shaft was equipped with four agitation paddles, two at the bottom and two at the middle of the biodigester. The shaft is also equipped with a speed reducer connected to an electric motor with 24 V power, which was activated by a timer for 15 minutes every hour, totaling twenty agitator activations per day, in order to promote biomass homogeneity.

The internal coils served to conduct the heated water from the boiler into the reactor, promoting heat exchange between the water and the biomass and keeping the internal temperature constant at 29 ± 0.5 °C. The electric pump was responsible for recirculating the water and was activated by the control panel, integrated with the temperature sensor, which activated the recirculation when the temperature sensor averaged temperatures below 28.5 °C and deactivated it when the sensor measured 29.5 °C.

4.5 BIOGAS ANALYSIS

The biogas flow rate was measured using a natural gas meter manufactured by LAO Indùstria® model G1, with maximum and minimum hourly flow rates varying between 2.300 and 0.020 m^3 , respectively. The biogas produced daily was stored in a plastic (vinyl) flask with a capacity of 1 m^3 in order to accumulate enough volume to sensitize the gas meter, since the biogas flow rate from the biodigester is lower than the meter's minimum reading (< 20 $L.h^{-1}$). When the balloon was fully inflated, which happened every 4 days, the biogas was expelled and its flow rate measured on the meter, a procedure that took around 90 minutes. The biogas production was then obtained, as well as the specific biogas production in liters per gram of reduced volatile solids ($L._{gSVr-1}$) and per reduced COD ($L.COD$)$^{-1}$

Qualitative analyses of the biogas were carried out using the LANDTEC GEM™ 5000 portable analyzer, three times a week, checking the levels of methane ($_{CH4}$), carbon dioxide ($_{CO2)}$ and hydrogen sulphide ($_{H2S}$). For greater credibility of the results obtained by the automatic analyzer, biogas analyses were carried out in the GC (described in section 4.6.2 of the methodology), which showed values very close to those obtained by the automatic analyzer, thus validating the data obtained.

4.6 BIOCHEMICAL METHANE POTENTIAL (BMP) TEST

The BMP (*Biochemical Methane Potential) test* consisted of evaluating the specific production of methane from food waste. This test is usually used to assess the ability of different substrates to produce methane under optimal conditions (OWEN et al., 1979; ANGELIDAKI et al., 2009; HAMILTON, 2012).

For the experiment, bench-top reactor flasks were used to check the potential for biogas generation. Each reactor consisted of a borosilicate flask with a volume of 250 mL, with polypropylene lids fitted with two registers, one for measuring biogas production and the other for discharging and collecting the biogas generated during the anaerobic digestion process, according to the methodology proposed by Alves (2008). A pressure gauge was attached to one of the registers to measure the internal pressure, with a maximum pressure reading of 2.5 $kgf.cm^{-2}$ and a scale of 0.10 $kgf.cm^{-2}$, as shown in Figure 4.6.1.

Two treatments were carried out: treatment A (control) consisted only of the inoculum; treatment B consisted of a mixture of inoculum and substrate, both worked in triplicate, as suggested by Angelidaki et al. (2009), in order to be statistically significant. The reactor feeding scheme is shown in Table 8.

Table 8. Batch reactor feeding scheme

Reactors	**Substrate volume**	**Volume Inoculum**	**Inoculum:substrate**	**Solids Volatile**
	mL		gSVinoculum/gSVsubstrate	g/L
A	0	100	1:0	11,62
B	0,9	100	1:1	14,05
Inoculum	-	-	-	11,62
Substrate	-	-	-	127,19

After each reactor flask received its substrate sample and inoculum, the lids were hermetically sealed on the glass and a stream of nitrogen gas (N_2) was recirculated in the *headspace* of each reactor flask for around four minutes, as shown in Figure 4, in order to ensure anaerobiosis of the medium.

The food waste (substrate) used in the BMP test was the same as that used in the prototype continuous reactor, and the inoculum was collected directly from the prototype reactor, which operated the mesophilic anaerobic digestion of food waste. It was decided to use the inoculum from the prototype reactor since it is of the same nature as the substrate, thus allowing the acclimatization phase to be faster, as well as providing better conditions for microbial development. However, there is still a lot of controversy about the best inoculum for a given substrate, as well as the optimum ratio between them (ARAMAL et al., 2008; SILVA, 2014).

For treatment B, a 1:1 waste:inoculum ratio was stipulated to carry out the interaction between the food waste and the microorganisms, in order to study the degradation rate. In general, the appropriate waste:inoculum ratio should be a slightly larger amount of inoculum than substrate (SILVA, 2014).

All the reactors were manually stirred daily for about 3 minutes to ensure greater contact between the microorganisms and the substrate (ANGELIDAKI et al., 2009).

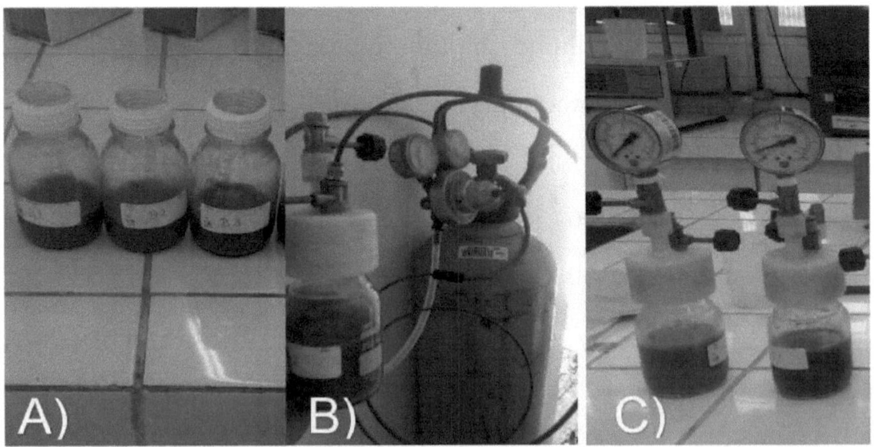

Figure 4. A) feeding the reactor flasks with their respective samples; B) recirculation of N2 inside the reactor flasks; C) coupling of the pressure gauge to the reactors to measure the internal pressures of the flasks. Source: Author.

The reactors were then incubated in a microbiological oven at a constant temperature of 32 ± 2 °C for a period of 30 days.

4.6.1 Collection and processing of biogas generation data from the Trial BMP

To calculate the volume of biogas generated, the following variables were monitored and recorded daily:

a) Internal pressure of the BMP reactors, in kgf.cm-2;

b) Temperature of the system in the oven, which was constant and equal to 32 ± 2 °C;

c) Local atmospheric pressure data, obtained from the *website of* the National Institute of Meteorology - INMET, from a weather station located near the UNIOESTE - Cascavel *campus*.

With this data, the internal pressure of the flasks was converted into the volume of biogas generated using equations (6) and (7). The volume values were corrected to the Normal Conditions of Temperature and Pressure - CNTP (MACIEL and JUCA, 2011):

$$\text{Generated between T e (T + 1)} = [\frac{PF \times VUF \times 22{,}41}{83{,}14 \times TF}] \times 1000 \qquad (6)$$

Where:

T = Time (days)

PF = Flask pressure in millibars (mbar)

VUF = Vial Usable Volume in liters (L)

TF = Flask Temperature in Kelvin (K)

Cumulative volume VA = Generated between T and (T + 1) + VGA (7)

Where:

VA = Cumulative Volume (mL)

VGA = Accumulated biogas volume from the previous day (mL)

In this way, the volume of biogas generated during the 30 days of anaerobic digestion at CNTP was obtained.

4.6.2 Gas chromatography for the biogas from the BMP test

In order to analyze the methane (CH_4) and carbon dioxide (CO_2) content of the biogas generated in the bench reactors, the biogas was collected on the tenth day of the experiment and stored in 37 mL gas vials developed by Construmaq Sâo Carlos (Figure 5). The collection procedure consisted of capturing the biogas from inside the reactors using a transfer kit consisting of a piston puller, a transfer tube and a probe, which allowed the biogas to be transferred from inside the reactors directly into the ampoule without contact with atmospheric air. This system is highly recommended for work with scarce samples, as is the case in this study, which used low quantities of organic matter in the reactors, consequently resulting in small volumes of biogas (CONSTRUMAQ, 2017).

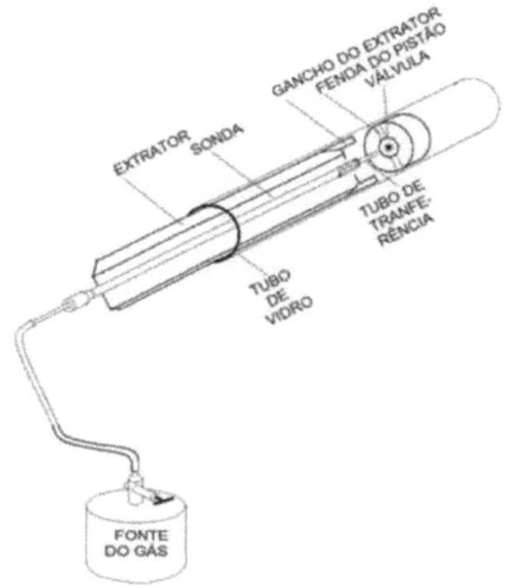

Figure 5 - Gas tube and biogas transfer kit.

Source: CONSTRUMAQ, 2016.

After capture, aliquots of the gas were collected in a syringe with a locking device (Sigma®) directly into the probe of the gas vials. The constituents of the biogas (carbon dioxide and methane) were determined by gas chromatography on a Shimadzu® 2010 system equipped with a Carboxen® 1010 plot capillary column (30 m x 0.53 mm x 0.30 µm). Argon was used as the carrier gas with a make-up air flow rate of 8 $mL.min^{-1}$. 500 µL of sample were injected and the injector temperature was set to 200°C . Detection was carried out using a

thermal conductivity (TCD) at a temperature of 230 °C. The furnace was programmed to operate at an initial temperature of 130 °C and heated to 135 °C at a rate of 46 $°C.min^{-1}$ for 6 minutes.

4.6.3 Statistical analysis and graph construction.

Statistical analysis was carried out using the R program, which consisted of simple analysis of variance and comparison of means using the Tukey method at 5% significance. The graphs were drawn up using the R program (version 2.13.0) and Microsoft Excel 2016.

5. RESULTS AND DISCUSSION

5.1 SUBSTRATE AND INOCULUM CHARACTERIZATION

The characteristics of the food waste used in the experiment are shown in Table 9, together with the data obtained by different authors who also used food waste as a substrate for biogas production. All the data was statistically validated, with a coefficient of variation below 0.5%.

Table 9. Comparative characterization of food waste.

Component	Zhang et al., 2007	Zhang et al., 2011	Zhang et al., 2013	Present Study
pH	NC	6,5	4,2	5,98
ST	30,9 %	18,1 %	23,1 %	15,28 %
SV	26,35 %	17,1 %	21,0 %	13,02 %
SV/ST	85,30 %	94 %	100 %	85,21 %
C/N	14,8	13,2	24,5	18,81
Total C	46,78 %	46,67 %	NC	44,4 %
Total N	3,16 %	3,54 %	NC	2,36%

NC: Nothing.

According to Zhang et al. (2007), there are no significant variations between the effluent and affluent of the anaerobic digestion process in terms of nutrient concentration. In fact, the microorganisms consume nutrients in their respective metabolisms, yet the micro and macro nutrients remained at fairly similar levels. In this study, a pH of 5.98 was obtained for the food waste, values within the limits found in the literature, just as Zhang et al. (2013) found a value of 4.2 for food waste, significantly acidic values that can inhibit the activity of microorganisms.

As for the organic fraction of food waste, as evidenced by the concentration of volatile solids, a percentage of 85.21 % was obtained, a value close to those found by Zhang et al. (2007) and Zhang et al. (2011), who found 85.30 % and 94 %, respectively. This therefore shows the presence of materials that can be converted into methane.

As for the C/N ratio, a value of 18.81 was found, which is slightly close to the lower limit desired for anaerobic digestion, given that the C/N ratio refers to digestion capacity, where an adequate ratio for the development of microorganisms would be in the range of 20 to 30 (VERNA, 2002). Even so, the value obtained in this study is close to those found in the literature for food waste. This is due to the carbon and nitrogen concentrations, which were very similar to those found by Zhang et al. (2007) and Zhang et al. (2011).

Table 10 shows the characteristics of the inoculum used to start the prototype biodigester.

Table 10. Characterization of the inoculum (swine manure).

Components	Values (based on dry weight)
pH	7,83
ST	25,5%
STF	18,4 %
SV/ST	72,15 %

According to the data shown in Table 10, the pig waste used as inoculum for starting the prototype biodigester had a pH of 7.83, which is close to the desired value for the anaerobic digestion process, which according to Silva (2014) should be close to 7, essential to ensure the inoculum is acclimatized when starting anaerobic reactors. In addition, considering the pH of the inoculum helped to neutralize the pH of the food waste, which was 5.68.

It is worth noting that the volatile content of the inoculum, in relation to the total solids, was 72.15%, indicating the presence of organic material that could be digested, since according to Decottignies et al. (2005), a waste can only be considered mineralized when it has volatile solids in the range of 10 to 17.4%.

Food waste is considered a promising organic substrate for anaerobic digestion due to its high potential for methane production (NEVES, OLIVEIRA and ALVES, 2009). However, inhibitions can occur when only biodigesting food waste for a long period of time. The reasons for the inhibition are the imbalance of nutrients in the fermentation, i.e. the trace elements (Zn, Fe, Mo, etc.) are insufficient, the macronutrients (Na, K, etc.) are excessive (ZHANG et al...), 2011; ZHANG, SU and TAN, 2013; EL-MASHAD and ZHANG, 2010) and the C/N ratio of food waste, at 18.81, is within the desired limits (SOSNOWSKI et al., 2003). Several researchers mention the advantages of co-digestion when compared to anaerobic digestions conducted with a single substrate (XU; LI, 2012; LUSTE et al., 2012; XIA et al., 2012; GIRAULT et al., 2012; RUGHOONUNDUN et al., 2012).

Zhang et al. (2013a) found that co-digestion of food waste with cattle manure not only increases the organic loading capacity, but also promotes methane yield in semi-continuous digestion. The high buffering capacity of co-digestion was observed due to the increased concentration of ammonia in the manure. In addition to ammonia, bicarbonate alkalinity is fundamental in anaerobic digestion, as it is a parameter that enables the neutralization of acids formed during the process and the buffering of pH, in the event of accumulations of volatile acids in the biodigester (CHERNICHARO, 2007).

The composition of organic substances, such as proteins and carbohydrates, is usually the part most easily available to microorganisms; on the other hand, lignocellulosic compounds are characterized by being difficult to degrade. According to the literature, food waste is rich in lipids, with around 5.0 g L^{-1} (ZHANG et al, 2013a). However, lipids are difficult to biodegrade and are considered a limiting factor in these processes. Some researchers attribute the failure of anaerobic digestion to the high concentration of lipids

(NEVES, OLIVEIRA E ALVES, 2006; OHA and MARTIN, 2010).

5.2 BIOCHEMICAL METHANE POTENTIAL TEST - BMP

The biochemical methane production potential of the food waste, which refers to the methane production yield, was evaluated by constructing generation rate and accumulated volume graphs. All methane volumes from the batch reactors were corrected to CNTP. Figure 6 shows graphs a and b, which refer to the biogas production of the treatments carried out in the BMP test.

a) Biogas generation rate

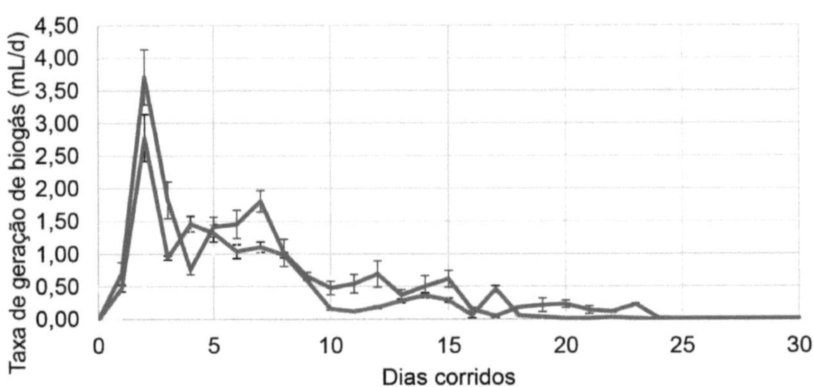

b) Cumulative biogas generation

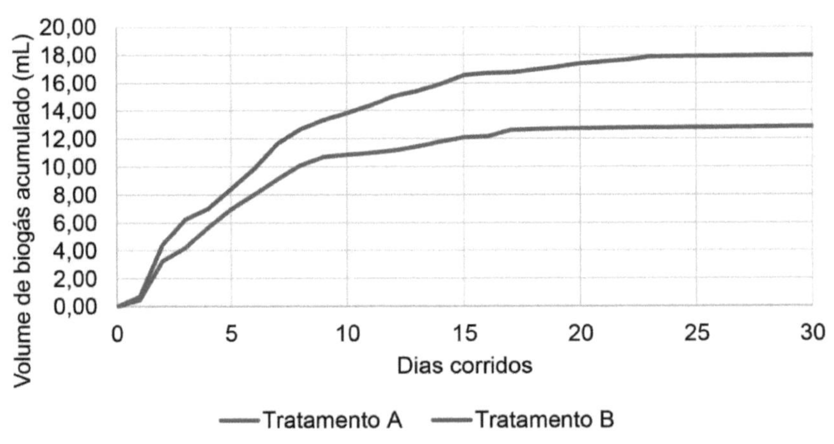

Figure 6: Biogas generation from food waste: a) generation rate; and b) accumulated generation.

Source: Author.

The results shown in Figure 6 represent the means and variances of treatments A and B, which had coefficients of variation (CV) of 0.28 and 1.79, respectively.

Figure 6 shows the speed of degradation of food waste, as well as the stability of biogas production around the 25th day, which according to Melo (2010) can be used as an indicator for the end of the BMP test. The drop in biogas production on the 25th day of operation of the BMP reactors was also observed by Silva (2015), who assessed the biodegradability of dairy waste and obtained the highest biogas generation on the 15th day of operation of the reactors, showing that the BMP test was able to generate consistent results for up to 30 days.

Figure 6 shows a peak in biogas production in the first few days of the test. This is due to the anaerobic conditions at the start of the experiment, which was achieved by recirculating N2 inside the reactor flasks, making the environment favorable for the development of anaerobic microorganisms and conducive to methanogenic activity (FIRMO, 2013). These peaks can also be explained by the presence of easily biodegradable soluble substances and the high volume of inoculum, which allowed for faster hydrolysis, consequently making the substrate quickly available to the methanogens, which in turn converted it into methane (PARAWIRA et al., 2004; FERREIRA, 2013; SILVA, 2014).

Angelidaki et al. (2009) also point out that particle size is an important parameter in the rate of biogas production, because reducing the particle size of the sample contributes to greater reactivity of the sample and speeds up the degradation time, which are fundamental for studies on a laboratory scale, as was the case in this experiment. In order to achieve homogeneity in the waste sample, the waste was ground up, thus increasing the surface area of the particles and promoting greater biogas production.

Treatment B (food waste + inoculum) contained approximately 20% more volatile solids than treatment A (pure inoculum), as shown in Table 8, yet the initial generation peak of both treatments was very similar, indicating that there were no inhibitions or adversities to the development of the microorganisms in the system. In other words, it can be considered an indication that the inoculum was able to acclimatize in the reactors and, above all, consume the available substrate (SILVA, 2014).

As for methane production, treatment B had an average 19.6% higher than treatment A, which was due to the greater availability of organic material and the interactions between the inoculum and the substrate, because where there is a greater concentration of organic material remaining in the environment, the stabilization (or biodegradation) rate is faster and, therefore, there is a greater production of biogas (VON SPERLING, 1996).

In general, the reactors showed results very similar to those found in the literature, with generation peaks at the start of the experiment and, over the course of the test, the curves began to show a tendency towards stable methane production. As in the tests carried out by Alves (2008) operating BMP reactors with MSW of different ages, who obtained the highest biogas production by the 5th day in all the tests. Firmo (2013) also presented very similar results when biodegrading food waste in BMP reactors, identifying the highest rates of biogas generation up to the 10th day of testing. Von Sperling (1996) explains that as the amount of organic matter in the reactor decreases, the rates of waste degradation and methane generation become lower.

On average, treatments A and B showed a methane concentration in the biogas of 68.74% and 57.48%, respectively. These values are within the desirable range, since, according to data found in the literature, methane levels from the anaerobic digestion of food waste vary between 50 and 80%, and in particular cases from 30% (JUCA et al., 2005; ALVES, 2008). The higher methane content for treatment A can be explained by the fact that it consisted of pure inoculum from an already active reactor containing an established methanogenic community; as for the methane content of treatment B (57.48%), it is explained by the fact that there was a greater generation of carbon dioxide in the fermentation phase of anaerobic digestion, which considerably increased the volume of biogas containing carbon dioxide in the mixture, consequently diluting the methane concentration (AMANI et al., 2010; ZHANG et al., 2014).

Silva (2014) also reports that when biodigesting food waste alone, without the aid of an inoculum, biogas is often produced, but with very low methane levels and sometimes even zero.

Table 11 shows the main response parameters of the BMP test.

Table 11. Characterization of the main parameters of the BMP test.

Parameter	Unit	Immediately A		Immediately B	
		Entry	Exit	Entry	Exit
pH		7,84	8,2	7,73	8,26
Total Solids	g.L1	18,57	17,92	24,04	17,696
Solid Volatile	g.L1	11,62	10,968	14,05	10,736
SV/ST	%	62,57		58,44	
COD	g O2.L-1	99,465	18,911	101,015	18,550
BMPbiogas	Nm^3 .kg DQOr-1	0,1596		0,2176	
	Nm^3 .kgSVr-1	0,419		0,541	

BMPmethane	Nm^3 .kg CODr^{-1}	0,1029	0,1251
	Nm^3 .kgSVr^{-1}	0,241	0,311
Methane/Biogàs	%	68,74 ± 1,36	57,48 ± 0,62
Reductions	% COD	80,98	81,63
	% SV	5,61	23,58

The data presented in Table 11 was statistically validated, and both showed a coefficient of variation of deviation from the mean of less than 0.3%.

The pH for both treatments showed values within the desirable range in the tributary, as presented by Bidone and Pivonelli (1999), who point out that microorganisms grow better at pHs close to neutrality; on the other hand, values outside the limits of 6.3 and 7.8 can inhibit the growth of some microorganisms. The inoculum (treatment A) had a pH of 7.84, just above the upper limit desirable for anaerobic digestion. Even so, no inhibitions were found in microbial activity, as evidenced by the production of methane in all replicates of the control reactors. However, after 30 days of biodigestion, the effluents from treatments A and B showed values significantly above the desirable limit, 8.2 and 8.26, respectively.

Crovador (2014) explains that the increase in pH at the end of the anaerobic process can often be due to the storage of biogas in the reactor's *headspace,* which promotes reactions of the soluble carbon dioxide with the reactor's aqueous solution. According to Henry's Law, which governs the solubility of gases, along with Dalton's Law, CO_2 remains in equilibrium varying between the gaseous and aqueous forms. With the production of CO_2 in biodegradation and the dynamics of Henry's law, this *headspace gas* solubilizes in the liquid phase of the biodigester (SCHIRMER et al., 2014), where it reacts with water to form bicarbonates, which implies an increase in alkalinity and pH (VON SPERLING, 2005).

As for the organic matter content, as evidenced by the volatile solids, it can be seen that treatment A has a moderate concentration (11.62 $g.L^{-1}$), because its organic matter had already been largely mineralized in the reactor that generated it. This factor also resulted in the low concentration of volatiles in treatment B (14.05 $g.L^{-1}$), because even with the addition of the substrate, in a 1:1 ratio, the concentration of volatiles rose very little. It should be borne in mind that the volume of substrate added was extremely low, as the ratio of 1:1 in $gSTV_{substrate}:gSTV_{inoculum}$ resulted in a volume of 0.009 L of substrate added in treatment B. However, the proportion of volatiles obtained of 62.57% and 58.44% for treatments A and B, respectively, is within the range considered optimal for biodegradation tests, due to the benefits promoted by microbial action (ALVES, 2008).

With regard to COD, the values found are consistent with the literature, where treatments A and B had a concentration of 99.45 to 101.01 $g\ O_2.L^{-1}$, respectively. Santos (2015), when biodigesting food waste in a pilot reactor, found a COD of 74.5 $g\ O_2.L^{-1}$ for the affluent sample containing food waste and inoculum (cattle manure). These values are slightly higher than those found by Ratanatamskul et al. (2014), who obtained a value of 232.795 $mg\ O_2.L^{-1}$ when characterizing food waste. Crovador (2014), when carrying out BMP tests with the organic fraction of MSW, obtained a COD for the affluent samples of 34.46 $g\ O_2.L^{-1}$.

It was possible to reduce the concentration of volatile solids in treatment A by 5.61% and in treatment B by 23.58%. This low reduction is due to the short hydraulic retention time of 30 days, which was not enough to mineralize the organic materials that are more difficult to degrade, such as the lignocellulosics present in food waste (CROVADOR, 2014). It should also be noted that in the effluents from both treatments, the concentration of volatile solids is very similar (around 10 to 11%), indicating that there are indeed recalcitrant materials that are difficult to degrade. However, Decottignies et al. (2005) points out that a waste can be considered stabilized when it has volatile solids between 10 and 17.4%; Kelly (2002) reports that waste can be considered stabilized when the volatile solids concentrations are less than 20%.

It can be said that there was digestion of the substrate, indicated by the reduction in COD, where it was possible to reduce it by 80.98 % in treatment A and 81.63 % in treatment B. This is considered to be a very significant removal due to the short hydraulic retention time (HRT) of 30 days, even though the process converted some of the solid organic matter into biogas. Santos (2015), operating anaerobic digestion in a psychrophilic regime of food waste in a pilot-scale reactor, achieved 84% COD removal and 94% SV removal.

With regard to the methane content in the production of biogas from the fermentation of food waste, the ratio of biogas and methane produced per unit of volatile solids reduced or consumed was used. Treatment A produced 0.419 Nm^3 .kgSVr-^{1} and 0.241 Nm^3 $_{CH4}$.kgSVr-^{1} , while treatment B produced 0.541 Nm^3 .kgSVr^^{1} and 0.311 Nm^3 $_{CH4}$.kgSVr^1 , values close to those found in the literature. Zhang et al. (2013) obtained in their research a biogas production yield from food waste of 0.621 Nm^3 .kgSVr^{-1} and 0.410 Nm^3 $_{CH4}$.kgSVr^{-1} . Haider et al. (2015) co-digested food waste with rice husks and achieved a specific biogas yield of 0.557 Nm^3 .kgSVr^{-1} . Cho, Park and Chang (1995) achieved a value of 0.472 Nm^3 $_{CH4}$.kgSVr-^{1} in BMP tests using food waste. This range is very similar to that obtained by Hansen et al. (2004), who reached a value of 0.495 Nm^3 $_{CH4}$.kgSV^^{1} . Therefore, even though food waste has very different composition characteristics from region to region, it is still an interesting renewable source for methane production, since it has an excellent yield compared to the other commonly used substrates shown in Table 6.

Compared to the yield obtained by the prototype biodigester, which was 0.444 Nm^3 CH4.kgSVr-1, there was a slightly lower production, which shows the efficiency of the prototype biodigester under the stipulated operating conditions, as well as pointing out that the BMP test could have been carried out with a higher concentration of substrate volatiles.

5.3 ANAEROBIC BIODIGESTER PROTOTYPE

The anaerobic digestion of food waste was carried out in a prototype anaerobic biodigester, designed and built specifically to operate the anaerobic digestion of food waste, as described in section 4.4 of this work. Therefore, as this prototype anaerobic digester is a new model which had not been tested before, it was necessary to identify which organic load would be the most suitable for the system. Thus, from the start of the biodigester, 5 stages were needed to stabilize it, due to instabilities found in the system due to under- or over-utilization related to the organic load added.

It should be noted that the volumetric organic load parameter is a fundamental factor to be studied in anaerobic digesters, since the efficiency of the system is linked to the maximum organic load capacity that the biodigester can withstand. On the other hand, the application of excessively high loads in anaerobic digesters leads to the accumulation of volatile fatty acids in the medium. As a result, instabilities in the process and possible system failures can occur (FERREIRA, 2015). From this perspective, we sought to find the best volumetric organic load that the biodigester prototype could withstand, increasing it progressively in order to obtain the real efficiency and stability of the system.

5.3.1 Stabilization stage 1

Fermentation started in the prototype biodigester on 15/09/2014. Simultaneously with the addition of water and inoculum, the biodigester was fed with an organic load of 0.30 g SVL^{-1} rd^{-1} . This load, gradually increased over 56 days, reached 1.18 g SVL rd^{-1-1} , but the pH dropped from 7.25 to 5.90. According to Li et al. (2010), the drop can be attributed to the high rate of hydrolysis and the accumulation of volatile fatty acids, which stops biogas production when the pH is below 6.5. On November 5, 2014, with a pH of 5.8, the methane content was 50% and the carbon dioxide content 47%, a typical result of fermentations under volatile fatty acid stress. To correct the pH to a range suitable for anaerobic fermentation, from 6.5 to 7.2 (APPELS et al., 2011), 627 g of hydrated lime (1.54 g of lime per liter of fermenting substrate) diluted in water was added. One day after the addition of lime (18/11/2014), the pH rose to 7.5, restoring favorable conditions for microbial development.

5.3.2 Stabilization stage 2

After correcting the pH, the load was gradually increased again, with the aim of reaching the stability limit for the second time, based on the acidity/alkalinity ratio, and reaching the data collection period. The load increase took place from 18/11/2014 to 18/02/2015, the day on which the load began to be repeated to get rid of the effect of its increase, when it is necessary to operate the biodigester for at least a time equal to the hydraulic retention time (HRT), which at this stage was 103 days. At the end of this period, on 12/06/2015, with the AV/AT ratio equal to 0.04, the biodigester was found to be underused, i.e. the organic load applied was below its digestion capacity, since the appropriate AV/AT ratio for the anaerobic digestion process is in the range of 0.1 to 0.5, and values well below this limit represent low production of organic acids in the process, which are a substrate for methanogens, implying lower methane production.

5.3.3 Stabilization stage 3

Therefore, a new load increase period began on June 15, 2015, with the aim of using the biodigester's full capacity, which lasted until August 31, 2015, when the AV/AT ratio was 0.25. This value is considered adequate in terms of stability in the production and consumption of acids in anaerobic fermentation (SANCHEZ et al., 2005). According to the authors, the ratio should be between 0.1 and 0.5 (AMANI et al., 2010; CABBAI et al., 2013).

From August 31st, 2015 until November 3rd, 2015, waiting to escape the influence of the increased load, the feed was maintained at an estimated organic load of 1.27 $gSVL\wedge^{1}$ $.rd\wedge^{1}$. On November 3, 2015, the sample collection period began to evaluate the biodigester's performance. However, on 12/09/2015, with a pH of 6.84, the AV/AT ratio showed a value of 0.9, signaling an imminent collapse of the fermentation (SILVA,1977 and SANCHEZ et al., 2005). Feeding was stopped immediately to prevent the pH from falling and the process from failing. It was concluded that the biodigester was not capable of digesting this load and that a new attempt should be made to adapt it to a digestible load.

5.3.4 Stabilization stage 4

After December 9, 2015, in order to maintain microbial activity, the biodigester was fed with a smaller volume of substrate and only on December 11 and 26, 2016 and January 5 and 18, 2016. In addition to the reduction in frequency, there was also a reduction in the organic load, translated into a lower daily feed volume. To monitor the fermentation response under these conditions, some values of the AV/AT ratio were obtained during this period, shown in Figure 7. Eight observations were made arbitrarily throughout the week, during the recovery period.

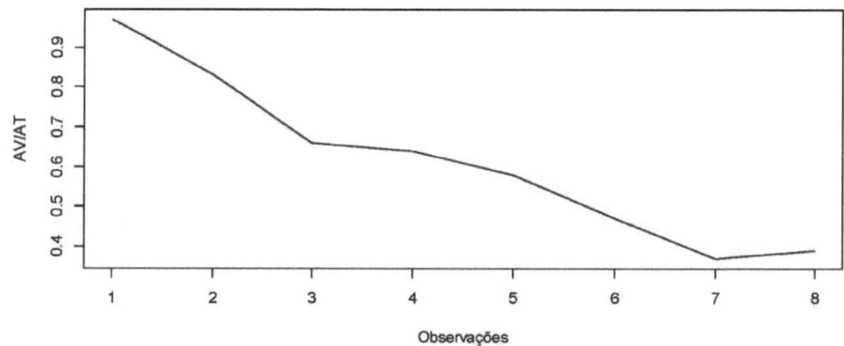

Figura 7. AV/AT ratio during the stabilization period of the anaerobic biodigester prototype.

Source: Author.

The feeding frequency was reduced by 86% to four operations. The strategy adopted led to a 31% reduction in the organic load added, increasing the alkalinity in the biodigester, as shown in Figure 7. In the 35 calendar days of the recovery period, the AV/AT ratio fell from 0.97 to 0.39, returning to the safe equilibrium conditions signaled by the variable. The biodigester was routinely fed from 2.ª to 6.ª .

5.3.5 Final stabilization stage

On 21/01/2016, with an AV/AT ratio of 0.36, a new period of regular feeding began, gradually increasing the volume of feed, taking into account the ratio between the acids and the buffer capacity of the fermenting biomass. The aim was to achieve sustainable stabilization and exploit the biodigester's maximum capacity to digest food waste. Biodigester stability was reached on 20/04/2016 with an organic load in COD of 0.82 $gL^{\wedge 1}$ $rd^{\wedge 1}$ and in volatile solids of 0.80 $gL^{\wedge 1}$ rd^ 1. From this date until 30/09/2016 was the time needed for the biodigester to come out of the effect of the increase in organic load and begin the data collection period. During this period, fermentation maintained an average AV/AT ratio of 0.37. The long period awaited for the start of data collection, which took place on 03/10/2016, was due to the high total solids content of the added substrate, which meant that the daily feed was only 2.5 L, for a 408 L biodigester, resulting in a TRH of 163 days.

5.3.6 Data collection phase

Once the stability of the biodigester was achieved and it was found that it was being fed with an adequate volumetric organic load, the data collection and analysis phase began. The main control parameters were checked: pH, temperature, alkalinity, AV/AT ratio, organic load, volatile solids, COD, specific biogas and methane production, and biogas components: CH_4, CO_2, H_2S and NH_3.

As described in the methodology of this study, 21 observations were made over 51 days, with the observations marked on the abscissa axis of the graphs referring to Mondays, Wednesdays and Fridays of the weeks in the data collection period, totaling 7 weeks of sampling.

Table 12 shows the results obtained for the parameters studied, which are discussed in the following sections.

Table 12. Characterizations of the biodigester's main parameters prototype.

Features	Unit	Value
Frequency	days	
food	week^{-1}	5
Biodigester volume	L	408
Mass of leftovers		
in b.u*	kg	1,125
Feed volume	Ld^{-1}	2,5
ST entry	%	15,285
ST outbound	%	1,950
ST reduction	%	87
pH output	-	8,02
AV/AT	-	0,491
Average temperature	°C	29,433
COD input	$g_{O2\ L-}^{\ 1}$	129,760
Outgoing COD	$g_{O2\ OL-}^{\ 1}$	21,778
SV entry	g.L^{-1}	130,293
SV out	g.L^{-1}	12,704
COD reduction	%	82,34
SV reduction	%	90,22
COD per liter	gL^{-1}	107,381
SV per liter	gL^{-1}	117,589
COD per day	gd^{-1}	268,453
SV per day	gd^{-1}	293,973
C. Organic (COD)	gDQO d^{-1}	324,400
C. Organic (SV)	gSV d^{-1}	325,733

C. Orgànica volumetric (COD)	$gDQOL^{-1}$ rd^{-1}	0,795
C. Orgànica volumetric (SV)	$gSVL^{-1}$ rd^{-1}	0,798
Biogàs per day	Ld^{-1}	220,470
Biogàspor vol. biodigester	LL^{-1} rd^{-1}	0,540
CH4 per day	Ld^{-1}	130,000
CH4 per volume biodigester	LL^{-1} rd^{-1}	0,319
CH4 by COD consumed	Lg^{-1} COD	0,506
CH4 by SV consumed	Lg-^{1} SVc	0,444

*b.u: wet base

Source: Author.

5.3.6.1. Biodigester stability control parameters

In order to understand the real functioning of the biodigester operating with food waste as a substrate, the main control parameters, already described in section 2 of this work, were analyzed. 21 observations were made over 51 days, starting on the 1st day of full operation of the biodigester with a TRH of 163 days, fed with 2.5 L of substrate.d^{-1} .

Figure 8 shows the pH analyses of the biodigester during the data collection period.

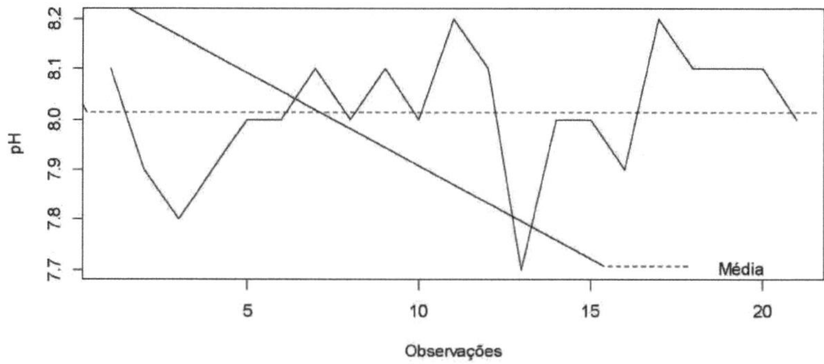

Figura 8. Variation in effluent pH during the data collection phase.

Source: Author.

The pH of the biodigester, on average 8.0, remained within expectations, slightly alkaline but still close to neutral. In some observations, values above 8.0 were noted, as shown in Figure 8. Such values can cause inhibitions and stresses in the system, especially with regard to methane production (BIDONE E POVINELLI, 1999). On the other hand, there were no disturbances in biogas and methane production, indicating that the microorganisms were able to withstand the pH variations.

This is justified by what happened in the BMP test in this research, where the pH of the effluent was in the range of 8.0 to 8.2, due to the interactions of carbon dioxide, which comes into equilibrium with the aqueous medium, resulting in the formation of bicarbonates. It can be considered that this also happened with this biodigester, where the biogas was stored in the storage flask and only discharged when the flask was inflated. Therefore, there was an accumulation of biogas inside the biodigester, allowing these reactions to occur (VON SPERLING, 2005; CROVADOR, 2014; SCHIRMER et al., 2014).

This pH behavior was also seen in the study by Ratanatamskul et al. (2014), operating the mesophilic anaerobic digestion of food waste in a prototype single-stage biodigester, similar to this study, fed with a substrate with a pH close to 4.5 and after the process the pH was slightly alkaline, around 7.9 to

8.0.

In anaerobic digesters, monitoring pH, volatile fatty acids and alkalinity is essential to ensure the efficiency of the process and its stability over long periods of hydraulic retention (RATANATAMSKUL et al., 2014). For this reason, the ratio of volatile acids to alkalinity (VA/TA) is considered one of the main parameters for controlling anaerobic processes (SANCHEZ et al., 2005; AMANI et al., 2010). If there is too much production of volatile acids during anaerobic digestion, due to the intense rate of hydrolysis and acidogenesis, a pH decrease can occur if the alkalinity of the system is insufficient. This can lead to partial or total inhibition of methanogenic microorganisms.

The stability of the anaerobic digestion process of food waste in the prototype anaerobic biodigester in this study, referenced in the AV/AT ratio, is shown in Figure 9. The AV/AT ratio is considered one of the main parameters for monitoring anaerobic digesters. It provides information on the stability of the process, indirectly representing whether there is an accumulation of acids in the system and whether the intermediate products of the process are being assimilated by the methanogenic *archaea* and, in fact, converted into methane.

The AV/AT ratio of the prototype biodigester is shown in Figure 9.

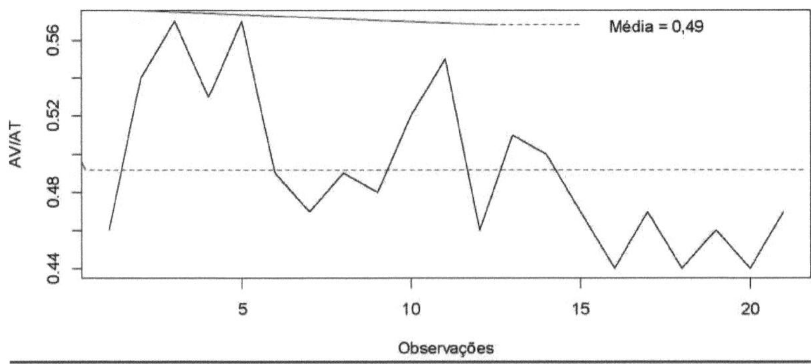

Figura 9. Variations in the AV/AT ratio during the data collection phase.

Source: Author.

The oscillations in the values of the AV/AT ratio were expected, due to the variations in the characteristics of the biodigester's feed substrate according to the menu of the popular restaurant, constituting different concentrations of elements in its composition. Even so, most of the observations remained within the desirable limits. According to Sànchez et al. (2005), in anaerobic processes the ideal value for the AV/AT ratio should be in the range of 0.1 to 0.5. In this case, the process had an average value of 0.49, close to the upper limit recommended by the authors. On the other hand, even with a significant concentration of volatile acids (VA) in the mixture, the pH of the biodigester remained close to neutral, denoting the strong buffering capacity of the mixture's alkalinity. Ratanatamskul et al. (2014), varying the hydraulic retention time and, consequently, the volumetric organic load, obtained values of 0.53, 0.61 and 0.79 in the AV/AT ratios for the respective TRHs of 27, 22 and 19 days. Ratanatamskul et al. (2015), when biodigesting food waste in a prototype two-phase anaerobic biodigester, found an AV/AT ratio of 0.59 ± 0.03. However, when he codigested food waste with sewage sludge, he obtained ratios of 0.32 ± 0.06, 0.33 ± 0.03, 035 ± 0.04 and 0.45 ± 0.03 with mixtures of food waste and sewage sludge in the proportions of 1:1, 3:1, 5:1, 7:1, respectively.

Therefore, considering the averages found in the literature, it can be said that the AV/AT ratio in this experiment is within the expected range for the monodigestion of food waste, which often lacks nutrients, causing a delay in the metabolism of methanogens and, consequently, a greater accumulation of intermediate products in the system (ZHANG et al., 2011; El-MASHAD & ZHANG, 2010).

5.3.6.2 Biodigester efficiency control parameters

In order to obtain information on the efficiency of the biodigester's organic load removal, the reductions in solids (ST and SV) and COD were checked. These parameters indirectly indicate the amount of organic material that can be biodegraded by microorganisms. Volatile solids are based on organic matter on a dry basis (GONÇALVES, 2012); COD measures the amount of oxygen needed to oxidize the organic fraction of a sample, which includes all material that can be oxidized, whether mineral or organic, thus providing an indirect reference on the organic load of a sample (VON SPERLING, 1996; SILVA, 2004).

The COD concentration over the data collection period of this study is shown in Figure 10.

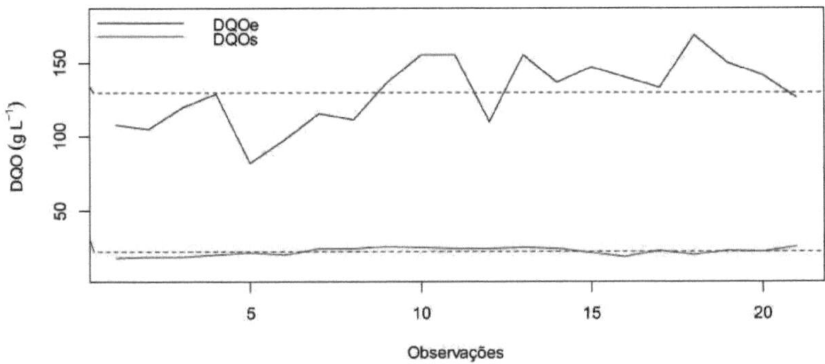

Figura 10. Variations in COD entering and leaving the prototype biodigester.

Source: Author.

The average COD concentration (affluent) obtained in this research, represented by the dashed line in Figure 10, for food waste was 129.76 g $_{O2}$.L^{-1} , a significantly high value compared to the values obtained by Ratanatamskul et al. (2014), who obtained a value of 23.27 g $_{O2.L}$-^{1} when characterizing food waste, and Ratanatamskul et al. (2015), who presented a COD of 23.57 g $_{O2}$.L^{-1} for food waste, in both cases the food waste came from the university restaurant of a Thai university.

However, it should be borne in mind that the characteristics of food waste vary greatly and are conditioned by eating habits, preparation techniques, social class, age group of consumers, etc. In Brazil, for example, the composition of food waste presents a great diversity of complex organic materials (hydrocarbons, lipids and fats), consequently presenting a greater organic load (SILVA, 2015). Santos (2015) found a value of 74.5 g $O2L^{-1}$ in the characterization of solid restaurant meal waste. Melo (2010), when characterizing the putrescible organic matter of solid urban waste from the Muribeca landfill (PE), found COD to be 49.56 g $_{O2}.L^{-1}$ and for sanitary waste COD was 29.87 g.$_{O2}.L^{-1}$.

As shown in Figure 10, it can be seen that anaerobic digestion of the organic fraction of food waste took place, as evidenced by the significant reduction in COD values (effluent). The reduction percentages can be seen in Figure 11, which also includes the SV reduction percentage.

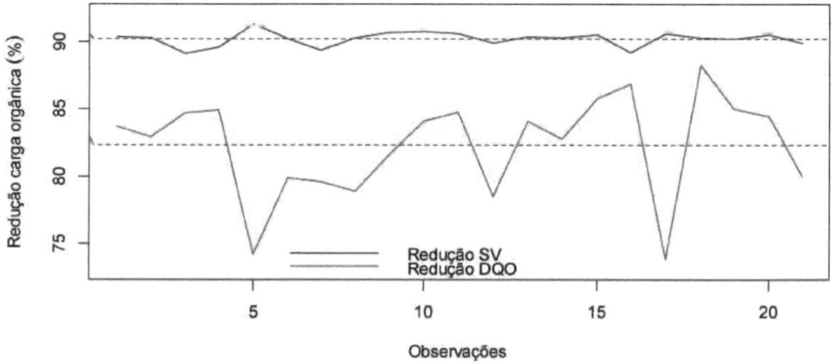

Figura 11. Percentages of COD and SV reductions.

Source: Author.

Figure 11 shows that the COD reductions were satisfactory, showing that the organic matter was degraded. On average, the reductions in COD were in the range of 82.34 % and in volatile solids 90.22 %.

Ratanatamskul et al. (2014) obtained the highest COD reduction of 73.71 % and SV reduction of 68.93 % when operating the anaerobic digestion of food waste in a mesophilic regime for a TRH of 27 days in a prototype anaerobic digester. The author attributed the low reductions to the short TRH used in the study, since when he carried out the experiment under the same conditions, varying only the TRH of 19 days, he obtained reductions of only

51.53% and 44.28% for COD and SV,

respectively. Santos (2015) operating anaerobic digestion of food waste obtained an average reduction in COD of 86% and in VS of 94%.

In this study, there was an efficient reduction in VS, in addition to the constancy of the values, which were always close to the average; in COD, there were significant amplitudes, with the highest reduction being 88.3% and the lowest 73.8%. Gonçalves (2012) and Neto (2015) report that the reduction in volatile solids only refers to organic matter on a dry basis and does not actually represent the conversion of organic matter into biogas and methane. In other words, the reduction percentages are not proportional to the amount of biogas produced. However, Ratanatamskul et al. (2014) report that with the conversion of organic matter into biogas, solids concentrations would also be reduced, which could be used as an indication of biogas production.

5.3.6.3 Biogas and methane production in the prototype biodigester

The efficiency of the anaerobic digestion process was also evaluated in terms of biogas production, both quantitatively and qualitatively. The biogas and methane yields obtained in this experiment were presented in reference to the reduction of SV and COD, since the experiment did not allow an analysis of daily biogas generation due to the need for biogas flow rates above 0.02 $m^3 .h^{-1}$ to sensitize the meter. Therefore, as explained in section 4.4 of this work, the biogas was first accumulated in the flask and when the flask was full, the biogas was expelled and the production measured. In this way, the average daily biogas production value was obtained, which was equal to 220 $L.d^{-1}$.

Therefore, from the daily production of biogas, methane production was obtained, shown in Figures 12 and 13, represented by the qualitative variations in methane levels as a function of the consumption of SV and COD, which were evaluated on the respective days of the observations made during the data collection period. Figures 12 and 13 show the specific methane production in relation to the reduction in SV and COD.

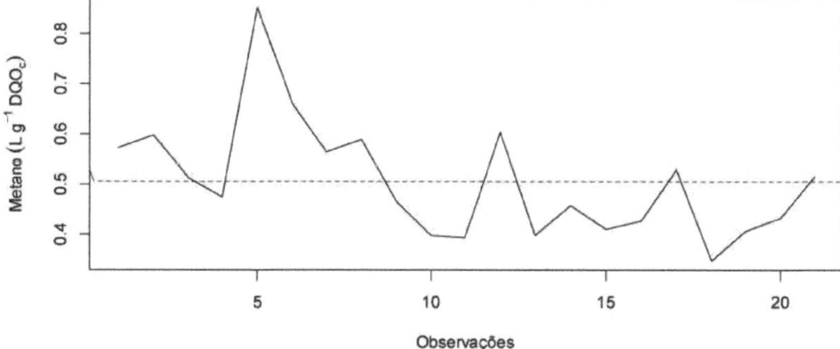

Figura 12. Variation in methane production as a function of COD reduction.

Source: Author.

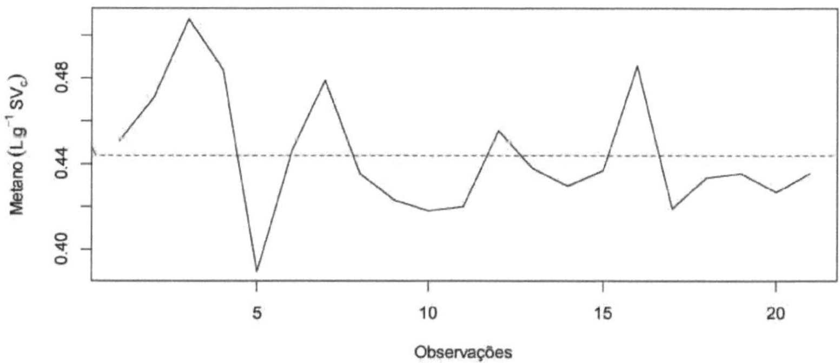

Figura 13. Variation in methane production as a function of SV reduction.

Source: Author

As shown in Figures 12 and 13, on average, the volumetric production of methane from reduced COD was 0.506 ± 0.11 L $_{CH4}$.gDQOr^{-1} and from reduced SV was 0.444 ± 0.02 L $_{CH4}$.gSVr^{-1} , values higher than those found in the BMP test in this research, indicating, even if indirectly, that the prototype biodigester was efficient in converting food waste into methane.

The values are also higher than those found in the literature, for example, Ratanatamskul et al. (2014) found indices of 0.077, 0.099 and 0.118 L

$_{CH4}$.gDQOr^{-1} and 0.237, 0.379 and 0.457 L $_{CH4}$.gSVr^{1} for TRHs of 27, 22 and 19 days, respectively.

With the anaerobic digestion of potato waste, Parawira et al. (2004) obtained a maximum methane production of 0.32 L $_{CH4}$.gSVr^{-1} . Kuczman (2007), when biodigesting cassava waste in a horizontal flow filter with bamboo as a support medium, obtained a maximum of 0.817 Lbiogas.kgDQOr^{-1} .

El-Mashad and Zhang (2010) found indices of 0.256 and 0.353 L $_{CH4}$.gSVr^{-1} , for TRHs of 20 and 30 days, respectively, operating the anaerobic digestion of food waste in batch reactors. Yong et al. (2015), when co-digesting food waste with straw, obtained the highest yield of 0.380 L $_{CH4}$.gSVr^{-1} , stating that the methane yield tended to increase as the proportion of food waste in the mixture increased. Even so, they report that compared to digestion of feed waste alone, codigestion with straw allowed methane production to increase by 39 %.

Mata-Alvarez et al. (1992) carried out tests in a continuous complete-mix biodigester, feeding it with fruit and vegetable waste, under a 20-day TRH, and found a methane yield of 0.47 L $_{CH4}$.gSVr^{-1} ; VERRIER et al. (1987) using a similar type of biodigester, with a TRH of 23 days, obtained a methane yield of 0.37 L $_{CH4}$.gSVr^{-1} .

Santos (2015), when biodigesting solid restaurant meal waste, obtained a maximum potential of 0.590 L $_{CH4}$.gSVr^{-1} . This production was slightly higher than that found in this study. The author notes that after the period of acclimatization of the inoculum, methane levels tend to rise. As explained by Koch, Helmreich and Drewes (2015), who found values in the range of 0.400 L $_{CH4.gSVr-}$^{1} after the biodigester was acclimatized with the inoculum.

Ferreira (2015) found an average methane production yield of 0.317 L CH4.gSV-1 in Phase 1 of the anaerobic digestion of food waste in a complete mixture biodigester, and reported this index as high, justifying it by the high activity of methanogens, often attributed to the microorganisms already established in the inoculum used.

However, it is clear that comparisons between the methane yields of the studies found in the literature are quite difficult due to variations in experimental conditions, which often have a direct influence on methane yields (PARAWIRA et al., 2004).

5.3.6.4 Biogas characterization: determination of CH4, CO_2 and H_2S.

The biogas from the prototype biodigester showed concentrations of methane (CH_4), carbon dioxide (CO_2) and hydrogen sulphide (H_2S), on average, at 58.86 %, 40.73 % and 835.25 ppm, respectively. There were no significant variations between the observations, as shown in Figure 14, which shows 8 analyses carried out arbitrarily during the data collection stage. The methane and carbon dioxide values are shown in percentages and the hydrogen sulphide content is shown in ppmv.

Figura 14. Concentration of the constituent gases in the biogas from the prototype biodigester.

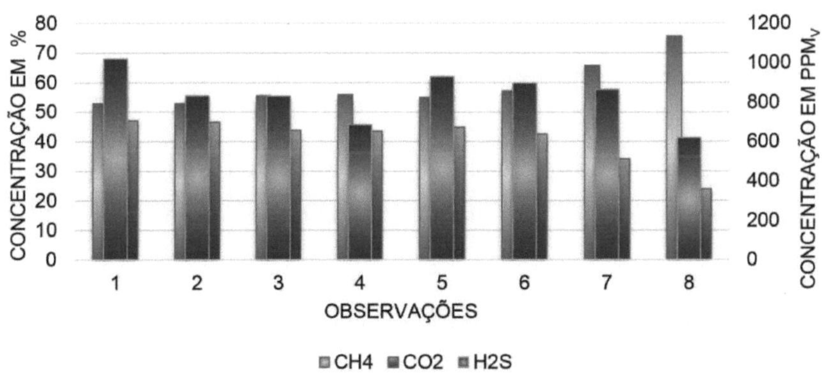

Source: Author.

Determining the concentrations of methane and other constituents of the biogas is an efficient tool for predicting possible stresses in the system, often due to thermodynamic limitations and accumulations of hydrogen and acetates that can cause a reduction or complete inhibition of the activity of the microorganisms involved in the process, especially methanogenic *archaea*, causing a drop in methane concentration and a reduction in the efficiency of organic matter conversion (CHERNICHARO, 2007).

Yong et al. (2015) found stable methane concentration ranges from 50 to 70% during the operating period of the anaerobic digestion of food waste with straw, in line with the methane levels found in this research, which ranged from 52.9 to 75.7%. Ratanatamskul et al. (2015) also found methane levels in the range of 50.2 to 60.4% in the anaerobic digestion of food waste in a prototype two-phase biodigester. El-Mashad and Zhang (2010) found methane levels in the range of 56.5 to 69.2 % when operating the anaerobic digestion of food waste with cattle manure. Therefore, it can be said that the prototype biodigester operated efficiently during the data collection stage, as the values found in this research are in agreement with the data obtained from the literature.

In contrast, Firmo (2013) obtained a maximum CH4 concentration for the anaerobic digestion of food waste in the range of 46.50%. Neto (2015), operating the anaerobic digestion of food waste in a prototype biodigester, achieved levels in the range of 28% CH_4, 71% CO_2, 300 ppm H_2S, values significantly lower than those obtained in this research, which confirms the efficiency of the prototype biodigester in converting organic matter into methane.

Hydrogen sulphide (H_2S), a toxic gas responsible for the bad smell of biogas, detectable at concentrations of 0.005 ppm (GOSTELOW, PARSONS and STUETZ, 2001), also has a high corrosive potential, showed average values in the biogas from the prototype biodigester of 835.25 ppm, with observations ranging from 619 ppm to 1021 ppm, typical values for biogas generated in domestic sewage treatment plants. Santos (2015), when digesting food waste in a prototype biodigester under psychrophilic conditions, found average H_2S

levels in the range of 832 ± 238 ppm and 776 ± 307, reporting that these values were close to the minimum limits for anaerobic digestion of food waste.

Desulphurization of biogas is deemed necessary for its use in energy cogeneration, where H_2S concentrations must be lower than 200 ppm in order to prolong the useful life of cogeneration system components (RYCKESBOSCH et al., 2011).

6. CONCLUSIONS

Food waste has proved to be a very promising substrate for biogas production, due to its wide availability in the country as well as its biodegradable characteristics, which have been verified in anaerobic digestion processes.

The physico-chemical characterizations of the food waste showed ideal values for the anaerobic digestion process, since it was possible to carry out the biodegradability tests with the material in natura, without the need to insert nutrient solutions. On average, the food waste had a volatile solids content of 85.21 % and a C/N ratio of 18.4, which according to the literature is favorable to the anaerobic digestion process. However, there were complications when digesting food waste in isolation, represented by oscillations in the AV/AT parameter, which indicated instabilities during the anaerobic digestion process in the prototype biodigester.

The Biochemical Methane Potential (BMP) test proved to be an efficient tool for verifying the biodegradability of food waste, generating satisfactory results in a short space of time (30 days). It was possible to obtain a methane yield in the range of 0.311 $m^3{}_{CH4}.kgSVr^{-1}$ and achieve organic matter reductions of around 23.58 % for VS and 81.63 % for COD in batch tests.

The prototype anaerobic digester was very efficient at removing organic matter from food waste, in the order of 82.34% COD and 90.22% SV. However, it took a few steps to find the right organic load to feed the prototype digester, represented by stability based on the AV/AT ratio, which had an average value of 0.49.

The methane and hydrogen sulphide contents obtained in the biogas from the prototype anaerobic digester were, on average, 58.86 % and 835.25 ppm_V, respectively. The methane yield obtained in the prototype biodigester was 0.506 m3 $_{CH4}.kgDQOr^{-1}$ and 0.444 m3 $_{CH4.kgSVr-}{}^{1}$.

It is concluded that food waste can be used in the anaerobic digestion process, and that the use of alternative substrates is advisable.

7. REFERENCES

ABRELPE - Brazilian Association of Public Cleaning and Special Waste Companies. **Panorama of Solid Waste in Brazil**. São Paulo, 2013.

ABRELPE - Brazilian Association of Public Cleaning and Special Waste Companies. **Panorama of Solid Waste in Brazil**. São Paulo, 2014.

ALVAREZ, J. M; MACE, S; LABRES, L. Anarobic digestion of solid wastes: An overview of research achievements and perspectives. **Bioresource Technology**. v. 74, p. 3-16, 2000.

ALVES, I.R. de F.S. **Experimental Analysis of the Potential for Biogas Generation in Urban Solid Waste.** 2008. 134f. Dissertation (Master in Civil Engineering). Federal University of Pernambuco - UFP, Recife - PE, 2008.

AMARAL, A.C. do; KUNZ, A.; STEINMETZ, R.L.R.; SCUSSIATO, L.A.; TAPPARO, D.C.; GASPARETO, T.C. Influence of solid-liquid separation strategy on biogas yield from a stratified swine production system. **Journal of Environmental Management**, v. 168, p. 229-235, 2016.

AMARAL, M. C. S.; FERREIRA, C. F. A.; LANGE, L. C.; AQUINO, S. F. Evaluation of the anaerobic biodegradability of landfill leachate. **Engenharia Sanitària e Ambiental,** v. 13, n. 1, p. 38-45, 2008.

AMANI, T.; NOSRATI, M.; SREEKRISHNAN, T.R. Anaerobic digestion from the viewpoint of microbiological, chemical, and operational aspects - a review. **Environmental Reviews**, n. 18, p. 255-278, 2010.

AMERICAN PUBLIC HEALTH ASSOCIATION - APHA. **Standard Methods for the Examination of Water and Wastewater.** American Public Health Association, American Water Works Association, Water Environmental Federation, 21.ed. Washington. 2005.

ANEEL. **Electricity atlas 2012**. Available <http://www.aneel.gov.br>. in: Accessed on December 10, 2015.

ANEEL. **Energy energy matrix.** Available at: <http://www2.aneel.gov.br/aplicacoes/capacidadebrasil/operacaocapacidadebrasil.cfm>. Accessed on: May 24, 2016.

ANEEL. **Normative Resolution No. 482 of April 17, 2012**. Establishes the general conditions for distributed microgeneration and mini-generation access to electricity distribution systems, the electricity compensation system, and other provisions.

ANGELIDAKI, I.; ALVES, M.; BOLZONELLA, D.; BORZACCONI, L.; CAMPOS, J.L.; GUWY, A.J.; KALYUZHNYI, S.; JENICEK, P.; VAN LIER, J.B. Defining the biomethane potential (BMP) of solid organic wastes and energy crops: a proposed protocol for batch assays. **Water Science & Technology**, v.59. n.5, p. 927-934, 2009.

APPELS, L.; ASSCHEB, A. V.; WILLEMSB, K.; DEGRÈVEA, J.; IMPEA, J. V.; DEWIL, R. Peracetic acid oxidation as an alternative pre-treatment for the anaerobic digestion of waste activated sludge. **Bioresource Technology,** v. 102, n. 5, p. 4124-4130, 2011.

ARAGON, G. **Proposal for a National Biogas and Biomethane Program** - PNBB. In: ARAGON, G (coord.). Brazilian Association of Biogas and Biomethane - ABIOGAS. Sâo Paulo, 2015. 69p.

BAZARA, X.; GALIMANY, F.; TORRES, R. Anaerobic digestion in the treatment of effluents and residual sludge. **Water Technology**, n. 233, p. 34-46, 2003.

BIDONE, F.R.A.; POVINELLI, J. **Basic concepts of solid waste**. Sâo Paulo: EESC/Universidade de Sâo Paulo. 120p. 1999.

BOUALLAGUI, H.; HAOUARI, O.; TOUHAMI, Y.; CHEICKH, R. B.; MAROUANI, L.; HAMDI, M. Effect of temperature on the performance of an anaerobic tubular reactor treating fruit and vegetable waste. **Process Biochemistry**, v. 39, p. 2143-2148, 2004.

BRAZIL, National Agency for Petroleum, Biofuels and Natural Gas. Resolution No. 8 of January 30, 2015. **Official Journal of the Union**, Brasilia, 2015. Provides for biomethane from organic agroforestry and commercial products and waste for use in vehicles (CNG) and residential and commercial installations.

BRAZIL. Decree No. 5.025, of March 30, 2004. Regulates item I and §§ 1, 2°, 3°, 4° and 5° of art. 3 of Law no. 10.438, of April 26, 2002.

Official Gazette of the Union, Brasilia, 2004. Provides for the Incentive Program for Alternative Sources of Electricity - PROINFA, first stage, and makes other provisions.

BRAZIL. Law n. 10.438, of April 26, 2002. Provides for the expansion of emergency electricity supply, extraordinary tariff recomposition, and other measures. **Official Gazette of the Union**, Brasilia, April, 2002.

BRAZIL. Law n. 12.305, of August 2, 2010. Establishes the National Solid Waste Policy; amends Law No. 9.605, of February 12, 1998; and makes other provisions. **Official Gazette of the Union**, Brasilia, Aug., 2010.

BRITTO, M. L. C. P. S. **Biogas emission rate and biodegradation parameters of urban solid waste at Aterro Metropolitano Centro**. 2006. 185 p. Dissertation (Professional Master's Degree in Environmental Management and Technologies in the Production Process) - Department of Environmental Engineering - DEA, UFBA, Salvador, 2006.

CABBAI, V.; BALLICO, M.; ANEGGI, E.; GOI, D. BMP tests of source selected OFMSW to evaluate anaerobic codigestion with sewage sludge. **Waste management,** v. 33, n. 7, p. 1626-1632, 2013.

CASSINI, S. T.; CHERNICHARO, C. A. L.; ANDREOLI, C. V.; FRANÇA, M.; BORGES, E. S. M.; GONÇALVES, R. F. Hydrolysis and anaerobic activity in sludge. In: CASSINI, S. T. (Coord.). **Anaerobic digestion of solid organic waste and use of biogas**. Rio de Janeiro: ABES, Projeto PROSAB, 2003.

210p.

CHANDRAN, P. K. P.; GANDHIRAJ, V. Co-digestion of the organic fraction of municipal solid waste with distillery effluent for enhanced methane generation. Proceedings Sardinia, 2009. Twelfth International Waste Management and Landfill Symposium S. Margherita di Pula, Cagliari, Italy: 5 - 9, October, 2009.

CHEN, G.; LIU, G.; YAN, B.; SHAN, R.; WANG, J.; LI, T.; WU, W. Experimental study of co-digestion of food waste and tal fescue for biogas production. **Renewable Energy**, v. 88, p. 273-279, 2016.

CHERNICHARO, C. A. L. **Anaerobic reactors.** Belo Horizonte: Department of Sanitary and Environmental Engineering - UFMG, v. 5, n. 2, 380 p., 2007.

CROVADOR, M.I.C. **Potential for generating biogas from the organic fraction of solid urban waste**. 119f. Dissertation (Master's Degree in Bioenergy). State University of the Center West - UNICENTRO, Irati, PR. 2014.

COELHO, S. T; VELAZQUEZ, S. M. S. G; SILVA, O. C. da; ABREU, F. C. de. Electricity generation from sewage treatment biogas using an 18 kW generator set. In: BRAZILIAN CONGRESS ON ENERGY PLANNING, 5, 2006. Brasilia. **Proceedings...** Sâo Paulo: Informaçâo Verbal, 2006. p. 1-12.

CORNELL COMPOSTING. **Estimating carbon content**. 1996. Available at: <http://compost.css.cornell.edu/calc/carbon.html>.

DECOTTIGNIES, V.; GALTIER, L.; LEFEBVRE, X; VILLERIO, T. **Comparison of analytical methods to determine the stability of municipal solid waste and related wastes.** In: Proceedings Sardinia, Tenth International Waste Management and Landfill Symposium, 2005.

DEUBLEIN, D.; STEINHAUSER, A. **Biogas from waste and renewable resources: An introduction.** Weinheim: Wiley-VCH Verlag Gmbh & Co. KGaA, 2008.

EL-MASHAD, H. M.; ZHANG, R. Biogas production from co-digestion of dairy manure and food waste. **Bioresource Technology**, v. 101, p. 4021-4028, 2010.

EPE - Energy Research Company. **Ten-Year Energy Expansion Plan 2024**. Ministry of Mines and Energy, Brasilia: MME/EPE. 2015.

FACTOR T. L.; ARAÙJO, J. A. C.; JÛNIOR, L. V. E. V. Pepper production in substrates and fertigation with biodigester effluent. Revista Brasileira de **Engenharia Agricola e Ambiental**, v.12, n.2, p.143-149, 2008.

FERREIRA, B.O. **Evaluation of a food waste methanization system with a view to the energetic use of biogas**. 2015. 124f. Dissertation (Master's Degree in Sanitation, Environment and Water Resources). School of Engineering - Federal University of Minas Gerais - UFMG, Belo Horizonte, MG. 2015.

FERREIRA, L. M. S. **Anaerobic biodigestion of dairy cattle waste with and without separation of the solid fraction.** 2013. 67f. Dissertation (Master in Animal Science) - Faculty of Agricultural and Veterinary Sciences - São Paulo State University - UNESP, Jaboticabal, SP, 2013.

FIRMO, A. L. B. **Numerical and experimental study of biogas generation from the biodegradation of solid urban waste.** 2013. 286 f. Thesis (PhD) - Postgraduate Program in Civil Engineering, Federal University of Pernambuco, UFPE, Recife, PE, 2013.

FONSECA, J. C. L.; SILVA, M. R. A.; BAUTITZ, I. R.; NOGUEIRA, R.F. P.; MARCHI, M. R. R. Evaluation of the analytical reliability of total organic carbon (TOC) determinations. **Eclética quimica**, v. 31, n. 3, p. 47, 52, 2006.

FORESTI, E. **Notes from the class Processes and Operations in Waste Treatment SHS-705.** Postgraduate course in Hydraulics and Sanitation, Sâo Carlos School of Engineering. 1993.

FOSTER-CARNEIRO, T; PÉREZ, M; ROMERO, L.I. Influence of total solid and inoculum contents on performance of anaerobic reactor treating food waste. **Bioresource Technology**, v. 99, p. 6994-7002, 2008.

GIRAULT, R.; BRIDOUX, G.; NAULEAU, F.; POULLAIN, C.; BUFFET, J.; PEU, P.; SADOWSKI, A.G.; BÉLINE, F. Anaerobic co-digestion of waste activated sludge and greasy sludge from flotation process: Batch versus CSTR experiments to investigate optimal design. **Bioresource Technology**, v. 105, p. 1-8, 2012.

GONÇALVES, C. D. C. **Modeling the FORSU anaerobic digestion process on an industrial scale**. 2012. 81 f. Dissertation (Master's degree in

Environmental Engineering) - Instiçâo de Engenharia, Arquitectura, Ciência e Tecnologia Tècnico Lisboa. Lisbon, Portugal, 2012.

GOSTELOW, P.; PARSONS, S.A.; STUETZ, R.M. Odour measurements for sewage treatment works. **Water Research**, V. 35, n. 3, p. 579-597, 2001.

GRIGATTI, M.; CIAVATTA, C.; GESSA, C. Evolution of organic matter from sewage sludge and garden trimming during composting. **Bioresource Technology**. v. 91 n. 2, p. 163-169, 2004.

GUNASEELAN, V. N. Biochemical methane potential of fruits and vegetables solid waste feedstocks. **Fuel and Energy Abstracts**, v. 36, n. 5, p. 403-403, 2004.

HAMILTON, D. W. **Anaerobic digestion of animal manures: methane production potential of wastes materials.** Oklahoma State University: Division of Agricultural Sciences and Natural Resources: BAE-1762. 2012, USA. 2012.

HASHIMOTO, A. G. Pretreatment of wheat straw for fermentation to methane. **Biotechnology and Bioengineering**. v. 28, p. 1857-66. 1986.

HAIDER, M. R.; ZESHAN; YOUSAF, S.; MALIK, R. N.; VISVANATHAN, C. Effect of mixing ratio of food waste and hice husk co-digestion and substrate to inoculum ration on biogas production. **Bioresource Technology**, v. 190, p. 451-457, 2015.

IBGE - **Brazilian Institute of Geography and Statistics**. National Basic Sanitation Survey 2010. Available at: < http://www.ibge.gov.br/home/estatistica/populacao/condicaodevida/pnsb2008/defaulttabpdf_man_res_sol.shtm> Accessed on: August 23, 2016.

IEA - Bioenergy. **Biogas upgrading technologies** - developments and innovations. Report of Task 37: Energy from biogas and landfill gas. International Energy Agency Technology Environment, Culham, Oxfordshire - UK. 2009. 20p.

IEA - Bioenergy. **Biogas upgrading and utilization**. Report of Task 24: Energy from biological conversion of organic waste. International Energy Agency Technology Environment, Culham, Oxfordshire - UK. 2001.20p.

Intergovernmental Panel on Climate Change - **IPCC**. Bruckner T., I.A. Bashmakov, Y. Mulugetta, H. Chum, A. de la Vega Navarro, J. Edmonds, A. Faaij, B. Fungtammasan, A. Garg, E. Hertwich, D. Honnery, D. Infield, M. Kainuma, S. Khennas, S. Kim, H.B. Nimir, K. Riahi, N. Strachan, R. Wiser, and X. Zhang, 2014: Energy Systems. In: Climate Change 2014: Mitigation of Climate Change. Contribution of Working Group III to the Fifth Assessment Report of the Intergovernmental Panel on Climate Change [Edenhofer, O., R. Pichs-Madruga, Y. Sokona, E. Farahani, S. Kadner, K. Seyboth, A. Adler, I. Baum, S. Brunner, P. Eickemeier, B. Kriemann, J. Savolainen, S. Schlomer, C. von Stechow, T. Zwickel and J.C. Minx (eds.)]. Cambridge University Press, Cambridge, United Kingdom and New York, NY, USA.

JUCA, J.F.T.; MACIEL, F.J.; MARIANO. M.O.H.; BRITO, A.R. **Technical Report on the Study of Energy Utilization from Biogas at the Muribeca Landfill**. Federal University of Pernambuco - Solid Waste Group, Recife - PE, 2005.

KAFLE, G. K.; CHEN, L. Comparison on batch anaerobic digestion of five different livestock manure and prediction of potential methane production (BMP) using different statistical models. **Waste Management**, v. 48, p. 492502, 2016.

KAFLE, G. K.; KIM, S. H.; SUNG, K. I.; Ensiling of fish industry waste for biogas production: A lab scale evaluation of biochemical methane potential (BMP) and kinetics. **Bioresource Technology**, v. 127, p. 326-336, 2013.

KARIM, K.; HOFFMANN, R.; KLASSON, T.; AL-DAHHAN, M.H. Anaerobic digestion of animal waste: Waste strength versus impact of mixing. **Bioresource Technology**, Amsterdam, v. 96, p.1771-1781, 2005.

KELLY, R. J. **Solid waste biodegradation enhancements and the evaluation of analytical methods used to predict waste stability**. Master Thesis (Master of Science in Environmental Science and Engineering) - Faculty of Virginia Polytechnic Institute and State University, Blacksburg-Virginia, 2002. 66p.

KEYMER, U.; REINHOLD, G. Grundsatze bei der projert planung. In: ROHSTOFFE,F.N. **Handreichchung biogasge winnung und-nutzung. Gülzow:** Institut Für Energetik Und Umwelt, p.182-209, 2006.

KHALID, A.; ARSHAD, M.; ANJUM, M.; MAHMOOD, T.; DAWSON, L. The anaerobic digestion of solid organic waste**. Waste Management**, v. 31, n. 8, p. 1737-1744, 2011.

KOCH, K.; HELMREICH, B.; DREWES, J. E. Co-digestion of food waste in municipal wastewater treatment plants: Effect of different mixtures on methane yield and hydrolysis rate constant. **Applied energy**, v. 137, p. 250 - 255. 2015.

LEMA, J. M.; MÉNDEZ, R. J.; *Anaerobic biological treatments*. Chapter: Contamination and environmental engineering. Vol. III, Contaminación de las aguas, Bueno, Julio, L. Sastre, H., Lavin, A. eds, F.I.C.Y.T, Oviedo, Espana.

1997.

LI, R. P.; WANG, K. S.; Li, X. J.; PANG, Y. Z. Characteristic and anaerobic digestion performances of kitchen wastes. **Renewable Energy**, v. 28, p. 7680, 2010.

LIM, S. J.; FOX, P. A kinetic evaluation of anaerobic treatment of swine wastewater at two temperatures in a temperate climate zone. **Bioresource Technology**, Amsterdam, v. 102, n.4, p.3724-3729, 2011.

LISSENS, G.; THOMSEN, A. B.; DE BAERE, L.; VERSTRAETE, W.; AHRING, B. K. Thermal wet oxidation improves anaerobic biodegradability of raw and digested biowaste. **Environmental Science and Technology**, v. 38, n. 12, p. 3418-3424, 2004.

LOPES, W. S.; LEITE, V. D.; PRASAD, S. Influence of inoculum on performance of anaerobic reactors for treating municipal solid waste. **Bioresource Technology**, v. 94, n. 3, p. 261-266, 2004.

LUSTE, S.; HEINONEN-TANSKIA, H.; LUOSTARINEN. S. Co-digestion of dairy cattle slurry and industrial meat-processing by-products - Effect of ultrasound and hygienization pre-treatments. **Bioresource Technology**, v. 104, p. 195-201, 2012.

MACHADO, L. L. N. **Technical aspects related to the generation of electricity from sewage sludge**. 2011. 109f. Dissertation (Master's Degree in Environmental Engineering). Federal University of Rio de Janeiro - UFRJ, Rio de Janeiro - RJ, 2011.

MACIEL, F. J.; JUCA, J. F. T. Evaluation of landfill gas production and emissions in a MSW large-scale experimental cell in Brazil. **Waste Management**, v.31, n.05, p.966-977, 2011.

MATA-ALVAREZ, J. Biomethanization of the organic fraction of municipal solid wastes. **IWA Publishing Company**, 2002.

MATA-ALVAREZ, J.; CECCHI, F.; LLABRÉS, P.; PAVAN, P. Anaerobic digestion of the Barcelona central food market organic wastes: experimental study. **Bioresour Technol**, v. 39, p. 39-48, 1992.

MAO, C.; FENG, Y.; WANG, X.; REN, G. Review on research achievements of biogas from anaerobic digestion. **Renewable and Sustainable Energy Reviews**, v. 45, p. 540-555, 2015.

MARTINS, F. M.; OLIVEIRA, P. A. V. de. Economic analysis of electricity generation from biogas in pig farming. Eng. Agricola, v.31, n.3, p.477-486, 2011.

MELO, E.S.R.L. de**. Analysis of the biodegradability of the materials that make up urban solid waste using BMP (Biochemical Methane Potential) tests.** 2010. 148f. Dissertation (Master in Civil Engineering). Federal University of Pernambuco - UFP, Recife, PE, 2010.

MINISTRY OF MINES AND ENERGY - MME. **Program to encourage alternative sources of electricity.** Available at: <http://www.mme.gov.br/programas/proinfa/menu/programa/tecnologias_contempladas.html>. Accessed on: September 1, 2016.

NELSON, D.W.; SOMMERS, L.E. **Methods of Soil analysis.** Part 3. Chemical Methods. Madison, WI. In: Bigham, J.M. Soil Science Society of America and American Society of Agronomy. Book Series No 5. p. 961-1010, 1996.

NETO, F. G. **Treatment of organic fractions of solid waste from restaurants in an anaerobic reactor in the municipality of Blumenau-SC**. 105f. Dissertation (Master's Degree in Environmental Engineering) Blumenau Regional University Foundation - FURB. Blumenau, SC. 2015.

NEVES, L.; RIBEIRO, R.; OLIVEIRA, R.; ALVES, M.M. Enhancement of Methane Production from Barley Waste. **Biomass and Bioenergy,** v. 30, n. 6, p. 599-603, 2006.

NEVES, L.; OLIVEIRA, R., ALVES, M.M. Anaerobic co-digestion of coffee waste and sewage sludge. **Waste Management v.** 26, n. 2, p. 176-181,2006.

OLIVEIRA, M. M. **Study of the inclusion of compartments in Canadian model reactors**. 2012. 119p. Dissertation (Master's Degree) - Postgraduate Program in Process Engineering, Federal University of Santa Maria - RS, 2012.

OHA, S.T.; MARTIN, A.D. Long chain fatty acids degradation in anaerobic digester: Thermodynamic equilibrium consideration. **Process Biochemistry**, v. 45, n. 3, p. 335-345, 2010.

PAES, R.F.C. **Characterization of the leachate produced at the Muribeca landfill - PE.** 2003. 150p. Dissertation (Master's Degree) - Center for Sciences and

Technology, Federal University of Campina Grande, Campina Grande - PB, 2003.

PARK, S.; LI, Y. Evaluation of methane production and macronutrient degradation in the anaerobic co-digestion of algae biomass residue and lipid waste. **Bioresource technology**, v. 111, p. 42-48, 2012.

PARAWIRA, W.; MURTO, M.; ZVAUYA, R.; MATTIASSON, B. Anaerobic batch digestion of solid potato waste alone and in combination with sugar beet leaves. **Renewable Energy**, v.29, n.11, p.1811-1823, 2004.

PECORA, V. **Implementation of a demonstration unit for generating electricity from biogas from the treatment of residential sewage at USP - Case Study**. 2006. 152 f. Dissertation (Master's Degree) - Interunits Program for Postgraduate Studies in Energy at USP, University of Sâo Paulo, Sâo Paulo, 2006.

PEROVANO, T. G; FORMIGONI, L. P. A. **Energy generation from by-products of sewage treatment.** 2011. 101f. CAPSTONE

(Bachelor of Environmental Engineering). University of Espirito Santo, Vitória

- ES. 2011.

PITK, P.; KAPARAJU, P.; PALATSI, J; AFFES, R.; VILU, R. Co-digestion of sewage sludge and sterilized solid slaugherhouse waste: Methane production efficiency and process limitations. **Bioresource Technology**. v. 134, p. 227232, 2013.

R DEVELOPMENT CORE TEAM. **R: a language and environment for statistical computing.** R Foundation for Statistical Computing, Vienna, Austria.

ISBN3-900051-07-0 , Available at :<//www.R-project.org/> Version

2.13.0. Accessed on: December 13, 2016.

RAPOSO, F.; DE LA RUBIA, M. A.; FERNANDEZ-CEGRi, V.; BORJA, R. Anaerobic digestion of solid organic substrates in batch mode: An overview relating to methane yields and experimental procedures. **Renewable and Sustainable Energy Reviews**, v. 16, n. 1, p. 861-877, 2012.

RAPOZO, F.; BANKS, C.J.; SIEGERT, I.; HEAVEN, S.; BORJA, R. Influence of inoculum to substrate ratio on the biochemical methane potential of maize in batch tests**. Process Biochemistry**, v. 41, n. 6, p. 1444-1450, 2006.

RATANATAMSKUL, C.; WATTANAYOMMANAPORN, O.; YAMAMOTO, K. An on-site prototype two-stage araerobic digester for co-digester of food waste and sewage sludge for biogas production from high-rise building. **International Biodeterioration & Biodegradation**, v. 102, p. 143-148, 2015.

RUGHOONUNDUNA, H.; MOHEEA, R.; HOLTZAPPLE, M.T. influence of carbon-to-nitrogen ratio on the mixed-acid fermentation of wastewater sludge and pretreated bagasse. **Bioresource Technology**, v. 112, p. 91-97, 2012.

RYCKEBOSCH, E; DROUILLON, M; VERVAEREN, H. Techniques for transformation of biogas to biomethane. **Biomass and Bioenergy**, v. 35, p. 1633-1645, 2011.

SALINO, P.J. **Energia eòlica no Brasil: uma comparação do proinfa e dos novos leiloes.** 84 f. TCC (Environmental Engineering). Polytechnic School, Federal University of Rio de Janeiro - UFRJ. Rio de Janeiro, 2011.

SANCHEZ, E.; BORJA, R.; TRAVIESO, L.; MARTIN, A.; COLMENAREJO, M.F. Effect of organic loading rate on the stability, operational parameters and performance of a secondary up flow anaerobic sludge bed reactor treating piggery waste. **Bioresource Technology**, Amsterdam, v. 96, n. 3, p. 335-344, 2005.

SANTOS, A. T. L.; HENRIQUE, N. S.; SHHILINDWEIN, J. A.; FERREIRA, E.; STACHIW, R. Utilization of the organic fraction of urban solid waste for the production of organic compost. **Revista Brasileira de Ciências da Amazônia**, v. 3, n. 1, p. 15-28, 2014.

SCHIRMER, W. N.; JUCA, J. F. T.; SCHULER, A. R. P.; HOLANDA, S.; JESUS, L. L. Methane production in anaerobic digestion of organic waste from Recife (Brazil) Landfill: 102 evaluation in refuse of different ages. **Brazilian Journal of Chemical Engineering**, v. 31, n. 02, p. 373-384, 2014.

SCHUCH, S.L. **Agroenergy condominium: potential for dissemination in agricultural activity**. 2012. 41 f. Dissertation (Master's Degree) Graduate Program in Agricultural Energy Engineering, State University of Western Paraná - UNIOESTE, Cascavel, 2012.

SILVA, G.A. **Estimation of biogas generation at the Joâo Pessoa metropolitan landfill using the BMP test.** 2012. 128f.

Master's Degree in Urban and Environmental Engineering, Federal University of Paraiba - UFP. 2012.

SILVA, G. H. **High-efficiency system for residential sewage treatment - a case study in Lagoa da Conceição**. Monograph. Civil Engineering degree program, Federal University of Santa Catarina, Florianópolis, 2004.

SILVA, G.B. **Evaluation of biogas production and methane generation from milk waste.** 2015. 18f. Capstone (Chemistry Technician). Univates University Center, Lajeado, RS. 2015.

SILVA, M.C.P. **Evaluation of anaerobic sludge and bovine manure as potential starter in anaerobic digesters for food waste.** 2014. 115f. Dissertation (Master's Degree) Postgraduate Program in Sanitation, Environment and Water Resources, Federal University of Minas Gerais - UFMG, Belo Horizonte, 2014.

SILVA, M.O.S.A. Physico-chemical analyses for the control of sewage treatment plants. Sâo Paulo: **CETESB**, 1977. 226p.

SGORLON, J. G.; RIZK, M. C.; BERGAMASCO, R.; TAVARES, C. R. G. Evaluation of COD and C/N ratio obtained in the anaerobic treatment of fruit and vegetable waste. **Acta Scientiarum Technology**, v. 33, n. 4, p. 421421,2011.

SOUZA, de M; SCHNEIDER, A. H. Opportunity of the biogas production chain for the state of Paranâ. In: SCHNEIDER, A. H. (coord.) **Observatórios Sistema Fiep** - SENAI. Curitiba/PR, 2016. 73p.

TOLMASQUIM, M. T. **Renewable Energy: Hydropower, Biomass, Wind, Solar, Ocean**. In: TOLMASQUIN, M. T. (coord.). Empresa de Pesquisa Energètica - EPE. Rio de Janeiro, 2016. 452p.

TCHOBANOGLOUS, G.; THEISEN, H.; VIGIL, S. Integrated Solid Waste Management: Engineering Principles and Management Issues. New York: Mcgraw-Hill, 1993. 978p.

VAN HAANDEL, A. C.; LETTINGA, G. Anaerobic sewage treatment. A manual for hot climate regions. Sâo Paulo, SP: **Mc. Graw Hill,** 1994. 199p.

VAZ, D. A.; FURIJO JR, A.; SOUZA, S. M. A. U.; SOARES, H. M. Anaerobic degradation kinetics of short-chain volatile acids in the presence of pentachlorophenol. In: **XIV SIMPÒSIO NACIONAL DE FERMENTAÇÔES,**

2003, Florianópolis. Complete Works, 2003, v. 1, p. 1-7.

VERNA, S. **Anaerobic digestion of biodegradable organics in municipal solid wastes.** 2002. 50p. Dissertation (M.Sc in Earth Resources Engineering). New York: Columbia University, 2002.

VERRIER, D.; RAY, F.; ALBAGNAC, G. Two-phase methanization of solid vegetable wastes. **Biol Wastes,** v. 22, p. 163-177, 1987.

VON SPERLING, M. **Basic principles of sewage treatment**. 1.ed. Belo Horizonte: Department of Sanitary and Environmental Engineering, Federal University of Minas Gerais, 1996. 211 p. (Principios do T ratamento Biològico de Aguas Residuárias, v. 2).

VON SPERLING, M. **Introduction to water quality and sewage treatment**. 3.ed. Belo Horizonte: Department of Sanitary and Environmental Engineering, Federal University of Minas Gerais, 2005. 452 p. (Principios do Tratamento Biològico de Aguas Residuárias, v. 1).

WEILAND, P. Biogas production: current state and perspectives. **Appl. Microbiol Biotechnol,** v. 85, p. 849-860, 2010.

WOON, K. S.; LO, I. M. C. A proposed framework of food waste collection and recycling for renewable biogas fuel production in Honk Kong. **Waste Management**, v. 47, p. 3-10, 2016.

XIA, Y.; MASSÉA, D.I.; MCALLISTERB, T.A.; KONGC, Y.; SEVIOURD, R.; BEAULIEU C. Identity and diversity of archaeal communities during anaerobic co-digestion of chicken feathers and other animal wastes. **Bioresource Technology**, v. 110, p. 111-119, 2012 .

XU, F.; LI, Y. Solid-state co-digestion of expired dog food and corn stover for methane production. **Bioresource Technology,** v..118, p. 219-226, 2012.

XU, C.; SHI, W.; HONG, J.; ZHANG, F.; CHEN, W. Life cycle assessment of food waste-based biogas generation. **Renewable and Sustainable Energy**

Reviews, v. 49, p. 169-177, 2015.

YONG, Z.; DONG, Y.; ZHANG, X.; TAN, T. Anaerobic co-digestion of food waste and straw for biogas production. **Renewable Energy**, v. 78, p. 527-530, 2015.

ZANETTE, A. L. Potential energy use of biogas in Brazil. 2009. 105 f. Dissertation (Master's in Energy Planning). Federal University of Rio de Janeiro - UFRJ, Rio de Janeiro, 2009.

ZANDONADI, H. S.; MAURÌCIO, A. A. Evaluation of the leftover-ingesta index of meals consumed by construction workers in the municipality of Cuiabà, MT. **Food Hygiene Magazine**. Sâo Paulo, v.26, n. 206/207, p. 64-70, 2012.

ZHANG, C.; SU, H.; BAEYENS, J.; TAN, T. Reviewing the anaerobic digestion of food waste for biogas production. **Renewable and Sustainable Energy Reviews**, v. 38, p. 383-392, 2014.

ZHANG, C.; SU, H.; TAN, T. Batch and semi-continuous anaerobic digestion of food waste in a dual solid-liquid system. **Bioresource Technology**, v. 145, p. 10-13, 2013a.

ZHANG, C.; XIAO, G.; PENG, L.; SU, H.; TAN, T. The anaerobic co-digestion of food waste and cattle manure. **Bioresource Technology**, v. 129, p. 170176, 2013b.

ZHANG, L.; LEEB, Y-W.; JAHNGA, D. Anaerobic co-digestion of food waste and piggery wastewater: Focusing on the role of trace elements. **Bioresource Technology**, v. 102, n. 8, p. 5048-5059, 2011.

ZHANG, R.; EL-MASHAD, H. M.; HARTMAN, K.; WANG, F.; LIU, G.; CHOATE, C.; GAMBLE, P. Characterization of food waste as feedstock for anaerobic digestion. **Bioresource Technology**, v. 98, p. 929-935, 2007.

Printed by Books on Demand GmbH, Norderstedt / Germany